ENVIRONMENTAL
CONTROL
IN
PETROLEUM
ENGINEERING

ENVIRONMENTAL CONTROL IN PETROLEUM ENGINEERING

JOHN C. REIS

Gulf Publishing Company
Houston, London, Paris, Zurich, Tokyo

ENVIRONMENTAL CONTROL
IN PETROLEUM ENGINEERING

Gulf Publishing Company
Book Division
P.O. Box 2608 □ Houston, Texas 77252-2608

10   9   8   7   6   5   4   3   2   1

Library of Congress Cataloging-in-Publication Data

Reis, John C.
    Environmental control in petroleum engineering /
John C. Reis.
        p.   cm.
    Includes bibliographical references and index.
    ISBN 0-88415-273-1 (alk. paper)
    1. Petroleum engineering—Environmental
aspects. 2. Pollution. I. Reis, John C. II. Title.
TD195.P4R45   1996
665.6—dc20   95-48462
                                              CIP

Printed on Acid-Free Paper (∞)

Contents

Acknowledgments

I would like to thank the many students who provided feedback on the course notes that eventually lead to this book. I would also like to thank Larry Henry for his thoughtful review of the manuscript. I gratefully acknowledge the donation of the reports by the American Petroleum Institute that are cited in this book.

Preface

With the rise of the environmental protection movement, the petroleum industry has placed greater emphasis on minimizing the environmental impact of its operations. Improved environmental protection requires better education and training of industry personnel. There is a tremendous amount of valuable information available on the environmental impact of petroleum operations and on ways to minimize that impact; however, this information is scattered among thousands of books, reports, and papers, making it difficult for industry personnel to obtain specific information on controlling the environmental effects of particular operations. This book assembles a substantial portion of this information into a single reference.

The book has been organized and written for a target audience having little or no training in the environmental issues facing the petroleum industry. The first chapter provides a brief overview of these issues. The second chapter focuses on the various aspects of drilling and production operations, while the third chapter discusses the specific impacts associated with them. Chapter 4 discusses ways in which toxic materials can be transported away from their release sites. (Actual waste transport modeling is a very complex topic and is beyond the scope of this book.) The fifth chapter presents ways to plan and manage activities that minimize or eliminate potential environmental impacts without severely disrupting operations. The sixth chapter discusses the treatment of drilling and production wastes to reduce their toxicity and/or volume before ultimate disposal. Chapter 7 presents disposal methods for various petroleum industry wastes. The final chapter reviews available technologies for remediating sites contaminated with petroleum wastes. A summary of major United States federal regulations, a list of major U.S. chemical waste exchanges, and discussions of sensitive habitats and offshore releases of oil are provided in the appendixes.

This book has evolved from course notes developed by the author for use in undergraduate and graduate classes. In preparing the book, the author has read thousands of pages of papers, reports, manuals, and books on the topic of environmental concerns facing the upstream petroleum industry. Although it is believed that this book is technically accurate, some errors and omissions have invariably occurred. There are many excellent papers and studies that are not included because the author did not become aware of them prior to publication of the book. The author welcomes constructive comments that may improve future editions.

ENVIRONMENTAL
CONTROL
IN
PETROLEUM
ENGINEERING

CHAPTER 1

Introduction to Environmental Control in the Petroleum Industry

The upstream petroleum industry, which conducts all exploration and production activities, provides essential petroleum products that are used for transportation fuels, electrical power generation, space heating, medicine, and petrochemicals. These uses of petroleum are major contributors to our present standard of living. The activities of finding and producing petroleum, however, can impact the environment, and the greatest impact arises from the release of wastes into the environment in concentrations that are not naturally found. These wastes include hydrocarbons, solids contaminated with hydrocarbons, water contaminated with a variety of dissolved and suspended solids, and a wide variety of chemicals. While some of these wastes can have significant adverse effects on the environment, some have little impact, and others are actually beneficial. In virtually all cases, the adverse impact can be minimized or eliminated through the implementation of proper waste management.

The most important steps in minimizing adverse environmental impact are for the industry to take a proactive approach to managing operations and become educated about those activities that can potentially harm the environment. The proactive approach involves adopting an attitude of environmental responsibility—not just to comply with regulations but to actually protect the environment while doing business.

1.1   OVERVIEW OF ENVIRONMENTAL ISSUES

Finding and producing oil and gas while minimizing adverse environmental impact requires an understanding of the complex issues facing the upstream petroleum industry. These issues concern operations that generate wastes, their potential influence on the environment, mechanisms and pathways for waste migration, effective ways to manage wastes, treatment methods to reduce their volume and/or toxicity, disposal methods, remediation methods for contaminated sites, and all applicable regulations.

1.1.1   Sources of Wastes

Wastes are generated from a variety of activities associated with petroleum production. These wastes fall into the general categories of produced water, drilling wastes, and associated wastes. Produced water accounts for about 98% of the total waste stream in the United States, with drilling fluids and cuttings accounting for the remaining 2%. Other associated wastes combined contribute a few tenths of a percent to the total waste volume (American Petroleum Institute, 1987). The total volume of produced water in the United States is roughly 21 billion barrels per year (Perry and Gigliello, 1990). A typical well can generate several barrels of fluid and cuttings per foot of hole drilled. In 1992, 115,903,000 feet of hole were drilled in the United States (American Petroleum Institute, 1993), yielding on the order of 300 million barrels of drilling waste.

Produced water virtually always contains impurities, and if present in sufficient concentrations, these impurities can adversely impact the environment. These impurities include dissolved solids (primarily salt and heavy metals), suspended and dissolved organic materials, formation solids, hydrogen sulfide, and carbon dioxide, and have a deficiency of oxygen (Stephenson, 1992). Produced water may also contain low levels of naturally occurring radioactive materials, or NORM (Gray, 1993). In addition to naturally occurring impurities, chemical additives like coagulants, corrosion inhibitors, emulsion breakers, biocides, dispersants, paraffin control agents, and scale inhibitors are often added to alter the chemistry of produced water. Water produced from waterflood projects may also contain acids, oxygen scavengers, surfactants, friction reducers, and scale dissolvers that were initially injected into the formation (Hudgins, 1992).

Drilling wastes include formation cuttings and drilling fluids. Water-based drilling fluids may contain viscosity control agents (e.g., clays), density control agents, (e.g., barium sulfate, or barite), deflocculants, (e.g., chrome-lignosulfonate or lignite), caustic (sodium hydroxide), corrosion inhibitors, biocides, lubricants, lost circulation materials, and formation compatibility agents. Oil-based drilling fluids also contain a base hydrocarbon and chemicals to maintain its water-in-oil emulsion. The most commonly used base hydrocarbon is diesel, followed by less toxic mineral and synthetic oils. Drilling fluids typically contain heavy metals like barium, chromium, cadmium, mercury, and lead. These metals can enter the system from materials added to the fluid or from naturally occurring minerals in the formations being drilled through. These metals, however, are not typically bioavailable. An extensive discussion of the environmental impacts of drilling wastes has been presented by Bleier et al. (1993).

Associated wastes are those other than produced water and drilling wastes. Associated wastes include the sludges and solids that collect in surface equipment and tank bottoms, pit wastes, water softener wastes, scrubber wastes, stimulation wastes from fracturing and acidizing, wastes from dehydration and sweetening of natural gas, transportation wastes, and contaminated soil from accidental spills and releases.

Another waste stream associated with the petroleum industry is air emissions. These emissions arise primarily from the operation of internal combustion engines. These engines are used to power drilling rigs, pumps, compressors, and other oilfield equipment. Other emissions arise from the operations of boilers, steam generators, natural gas dehydrators, and separators. Fugitive emissions from leaking valves and fittings can also release unacceptable quantities of volatile pollutants.

One common, but incorrect, perception of the petroleum exploration and production industry is that it is responsible for large-scale hydrocarbon contamination of the sea. The total amount of hydrocarbons that enter the sea is estimated to be 3.2 million metric tons per year. The individual contributions from the different sources of hydrocarbons is given in Table 1-1 (National Research Council, 1985). The primary source of hydrocarbon releases into the ocean is from transportation

Table 1-1
Sources of Hydrocarbon Inputs into the Sea

Source	Amount Introduced (metric tons/year)
Natural Sources	**0.25**
Marine seeps	(0.2)
Sediment erosion	(0.05)
Offshore Production	**0.05**
Transportation	**1.47**
Tanker operations	(0.7)
Dry-docking	(0.03)
Marine terminals	(0.02)
Bilge and fuel oils	(0.3)
Tanker accidents	(0.4)
Nontanker accidents	(0.02)
Atmospheric Transport	**0.3**
Municipal and Industrial	**1.18**
Municipal wastes	(0.7)
Refineries	(0.1)
Nonrefining industrial wastes	(0.2)
Urban runoff	(0.12)
River runoff	(0.04)
Ocean dumping	(0.02)
TOTAL	**3.2**

Source: from National Research Council, 1985.
Copyright © 1985, National Academy of Sciences.
Courtesy of National Academy Press, Washington, D.C.

by tankers. Oil production from offshore platforms contributes less than 2% of the total amount of oil entering the sea.

1.1.2   Environmental Impact of Wastes

The primary measure of the environmental impact of petroleum wastes is their toxicity to exposed organisms. The toxicity of a substance is most commonly reported as its concentration in water that results in the death of half of the exposed organisms within a given length of time. Exposure times for toxicity tests are typically 96 hours, although other times have been used. Common test organisms include mysid shrimp or sheepshead minnows for marine waters and fathead minnows or rainbow trout for fresh waters.

The concentration that is lethal to half of the exposed population during the test is called LC_{50}. High values of LC_{50} mean that high concentrations of the substance are required for lethal effects to be observed, and this indicates a low toxicity. A related measure of toxicity is the concentration at which half of the exposed organisms exhibit sublethal effects; this concentration is called EC_{50}. Another measure of toxicity is the *no observable effect concentration* (NOEC), the concentration below which no effects are observed.

The environmental impact of hydrocarbons in water varies considerably (National Research Council, 1985). The toxicity of aromatic hydrocarbons is relatively high, while that of straight-chain paraffins is relatively low. LC_{50} values for the most common aromatic hydrocarbons found in the petroleum industry (benzene, toluene, xylene, and ethylbenzene) are on the order of 10 ppm. Hydrocarbon concentrations of less than 1 mg/l in water have been shown to have a sublethal impacts on some marine organisms. High molecular weight paraffins, on the other hand, are essentially nontoxic. Chronic exposures of entire ecosystems to hydrocarbons, either from natural seeps or from petroleum facilities, have shown no long- or intermediate-term impact; the ecosystems have all recovered when the source of hydrocarbons was removed. No evidence of irrevocable damage to marine resources on a broad oceanic scale, by either chronic inputs or occasional major oil spills, has been observed. Although there are short-term impacts from major spills, the marine resources can and do recover.

Other effects of hydrocarbons include stunted plant growth if the hydrocarbon concentration in contaminated soil is above about 1% by weight. Lower concentrations, however, can enhance plant growth (Deuel, 1990). Hydrocarbons can also impact higher organisms that may become exposed following an accidental release. Marine animals that use hair or feathers for insulation can die of hypothermia if coated with oil. Coated animals can also ingest fatal quantities of hydrocarbons during washing and grooming activities.

The high dissolved salt concentration of most produced water can also impact the environment. Typical dissolved salt concentrations for produced water range between 50,000 and 150,000 ppm. By comparison, the salt concentration in seawater is about 35,000 ppm. Dissolved salt affects the ability of plants to absorb water and nutrients from soil. It can also alter the mechanical structure of the soil, which disrupts the transport of air and water to root systems. Water with dissolved salt concentrations below about 2,500 mg/l have minimal impact on most plants (Deuel, 1990). LC_{50} values for dissolved salt concentrations for freshwater organisms are on the order of 1,000 ppm (Mount et al., 1993).

The toxicity of drilling muds varies considerably, depending on their composition. Toxicities (LC_{50}) of water-based muds containing small percentages of hydrocarbons can be a few thousand ppm. The LC_{50}s of polymer muds, however, can exceed one million, which means that fewer than 50% of a test species will have died during the test period.

The toxicity of heavy metals found in the upstream petroleum industry varies widely. The toxicity of many heavy metals lies in their interference with the action of enzymes, which limits or stops normal biochemical processes in cells. General effects include damage to the liver, kidney, or reproductive, blood forming, or nervous systems. With some metals, these effects may also include mutations or tumors. Heavy metal concentrations allowed in drinking water vary for each metal, but are generally below about 0.01 mg/L. The heavy metals in offshore drilling fluid discharges normally combine quickly with the naturally abundant sulfates in seawater to form insoluble sulfates and precipitates that settle to the sea floor. This process renders the heavy metals inaccessible for bioaccumulation or consumption.

Nuclear radiation from NORM can disrupt cellular chemistry and alter the genetic structure of cells. In most cases, however, radiation exposure from NORM is significantly lower than that from other natural and man-made sources of radiation and does not represent a serious health hazard (Snavely, 1989).

The various chemicals used during production activities can also affect the environment. Their toxicities vary considerably, from highly toxic to essentially nontoxic. In most cases, however, the concentrations of chemicals actually encountered in the field are below toxic levels (Hudgins, 1992).

The primary environmental consequences of air pollutants are respiratory difficulties in humans and animals, damage to vegetation, and soil acidification. Releases of hydrogen sulfide, of course, can be fatal to those exposed.

1.1.3   Waste Migration

In most cases, the environmental impact of released wastes would be minimal if the wastes stayed at the point of release; unfortunately, most wastes migrate from their release points to affect a wider area. The migration pathway most often moves through groundwater along the local hydraulic gradient. For releases at sea, wastes will follow the prevailing winds and currents. For air emissions, the pollutants will follow the winds. Because migration spreads the wastes over a wider area, the local concentrations and toxicities at any location will be reduced by dilution.

1.1.4   Managing Wastes

The most effective way to minimize environmental impact from drilling and production activities is to develop and implement an effective waste management plan. Waste management plans identify the materials and wastes at a particular site and list the best way to manage, treat, and dispose of those wastes (Stilwell, 1991; American Petroleum Institute, 1989). A waste management plan should also include an environmental audit to determine whether existing activities are in compliance with relevant regulations (Guckian et al., 1993).

The effective management of each waste consists of a hierarchy of preferred steps. The first and usually most important step is to minimize the amount and/or toxicity of the waste that must be handled. This is done by maintaining careful control on chemical inventories, changing operations to minimize losses and leaks, modifying or replacing equipment to generate less waste, and changing the processes used to reduce or eliminate the generation of toxic wastes.

The next step in effective waste management is to reuse or recycle wastes. If wastes contain valuable components, those components can be segregated or separated from the remainder of the waste stream and recovered for use. Wastes that cannot be reused or recycled must then be treated and disposed of. A written waste management plan that completely describes the acceptable options for handling every waste generated at every site must be developed and effectively communicated to every employee involved with the wastes. Examples of how the waste management hierarchy can be implemented are given by Thurber (1992), Derkies and Souders (1993), and Savage (1993).

In most cases, the cost of eliminating all risks and hazards associated with wastes is economically prohibitive. Prudent management practices focus available resources on the activities that pose the greatest risk to both the economic health of the company and the environment. The risks associated with various waste management practices can be quantified and ranked through risk assessment studies (Sullivan, 1991). When properly managed, the risks and hazards of drilling and production operations can be reduced to low levels.

1.1.5   Waste Treatment Methods

Most wastes require some type of treatment before they can be disposed of. Waste treatment may include reducing the waste's total volume, lessening its toxicity, and/or altering its ability to migrate away from its disposal site. A variety of treatment methods are available for different types of wastes, although their costs vary significantly. The waste treatment method selected, however, must comply with all regulations, regardless of their cost.

One of the most important steps in waste treatment is to segregate or separate the wastes into their constituents, e.g., solid, aqueous, and hydrocarbon wastes. This isolates the most toxic component of the waste stream in a smaller volume and allows the less toxic components to be disposed of in less costly ways. Primary separation occurs with properly selected and operated equipment, e.g., shale shakers, separation tanks, and heater treaters. Separation can be improved by using hydrocyclones, filter presses, gas flotation systems, or decanting centrifuges (Wojtanowicz et al., 1987). In arid areas, evaporation and/or percolation can be used to dewater some wastes.

A number of methods are available for treating hydrocarbon-contaminated solids like drill cuttings, produced solids, or soil. Solids can be washed by agitation in a jet of high-velocity water, perhaps with an added surfactant. Solids can also be mixed with an oil-wet material such as coal or activated carbon, that absorbs the hydrocarbons and can be separated from the more dense solids by subsequent flotation. An emerging and promising technology for hydrocarbon removal from contaminated solids is *bioremediation.* Other treatment methods include distillation, solvent extraction, incineration, and critical/supercritical fluid extraction.

Nonhydrocarbon aqueous wastes can be treated by a number of methods, including ion exchange, precipitation, reverse osmosis, evaporation/distillation, biological processes, neutralization, and solidification. These processes can remove dissolved solids from water or encase them in other solids to prevent subsequent leaching following disposal.

1.1.6 Waste Disposal Methods

A number of disposal methods are available for petroleum industry wastes. The method used depends on the type, composition, and regulatory status of the waste.

The primary disposal method for aqueous wastes is to inject them into Class II wells. If the quality of wastewater meets or exceeds regulatory limits, permits to discharge it into surface waters may be obtained in some areas.

The primary disposal methods for solid wastes are to bury them or to spread them over the land surface. All free liquids normally must be removed prior to disposal, either by mechanical separation, evaporation, or the addition of solidifying agents. Land treatment of wastes may be prohibited if volatile and leachable fractions are present in the wastes. Disposal can occur either on or off-site. Underground injection of slurries has also been used for solids disposal in some areas.

1.1.7 Cleanup Methods for Contaminated Sites

The most appropriate cleanup method will depend on the contaminant and on the site characteristics. The most common contaminated sites are those that have spilled hydrocarbons in the soil and those containing old drilling fluids.

A number of methods can be used to clean up sites. Mobile hydrocarbons can be removed by drilling wells or digging trenches and pumping the hydrocarbons to the surface with groundwater for treatment. Volatile hydrocarbons can be removed by injecting air and/or pulling a vacuum to vaporize those components. The use of heat, surfactants, and bioremediation to remove subsurface hydrocarbons is being studied. Dissolved hydrocarbons in water and volatilized hydrocarbons in air can be removed by filtration or by absorption with activated carbon. In some cases, however, the contaminated material may need to be completely removed for offsite treatment and disposal.

1.1.8   Environmental Regulations

One of the most significant changes occurring in the operations of the upstream petroleum industry during the 1980s has been the need to minimize environmental impact. This change has been driven by an increase in the number of regulations governing drilling and production activities. Most of these regulations impose economic fines and possibly criminal penalties for violations. These regulations have significantly increased the cost of industry operations.

Major United States Environmental Regulations and Costs

A number of major environmental regulations affect the operation of petroleum exploration and production activities in the United States (Gilliland, 1993; Interstate Oil Compact Commission, 1990). Some of these regulations are briefly reviewed below; a more extensive discussion of the regulations is included in Appendix A.

The Resource Conservation and Recovery Act (RCRA), Subtitle C, regulates the storage, transport, treatment, and disposal of hazardous materials that are intended to be discarded, i.e., wastes. This regulation defines hazardous wastes as those that are specifically listed by name or those that are either highly reactive, corrosive, flammable, or toxic. Most, but not all, upstream petroleum industry wastes are exempt from this regulation.

The Safe Drinking Water Act was passed to protect underground sources of drinking water (USDW). This act regulates activities that may contaminate USDWs, particularly injection wells for both oil recovery and water disposal, as well as the plugging of abandoned wells. This act requires regular mechanical integrity testing of all injection wells.

The Clean Water Act prohibits the discharge of wastes, particularly oil, into surface waters or drainage features that may lead to surface waters. This act requires many operators to prepare *spill prevention control and countermeasure* (SPCC) plans to help minimize the impact of any spills.

The Clean Air Act regulates the emissions of air pollutants, including exhaust from internal combustion engines, fugitive emissions, and boiler emissions. This act specifies the types of emissions control equipment that must be used.

The Comprehensive Environmental Response, Compensation, and Liability Act (CERCLA or Superfund) was enacted to identify existing sites where hazardous wastes may impact human health. It established cleanup and claims procedures for affected parties. The Superfund Amendments and Reauthorization Act (SARA) requires that facilities storing hazardous materials keep a written inventory of those materials and provide them to local authorities. Crude oil is considered non-hazardous under this act, while many of the other RCRA exempt wastes are considered hazardous.

The potential costs of environmental regulations on the exploration and production of oil have been studied (Godec and Biglarbigi, 1991; Perkins, 1991). Depending on how these regulations are interpreted and implemented, the resulting loss of production may be as high as 50% of that without the environmental regulations. If the economic costs of these regulations in the U.S. is prorated over the existing production levels, the resulting costs would be a few dollars per barrel of oil produced.

1.2   A NEW ATTITUDE

We are all environmentalists. We all want a clean place to live. We all want clean water to drink. We all want clean air to breath. We all want to live in a world safe from toxic hazards. We all want to live in a world that is aesthetically pleasing. Yet, we also want the benefits of inexpensive energy. We want to be able to drive our cars, fly our planes, have electric lights and appliances in our homes, and keep our homes warm in the winter and cool in the summer. We want the medicines and plastics made from hydrocarbons. But often, the desire for a pristine environment and the benefits of inexpensive energy conflict. To drive our cars, we must find, produce, and transport crude oil. To maintain access to the benefits of inexpensive energy, we need a strong domestic petroleum industry.

There will always be the risk of environmental harm during exploration and production activities. There are risks associated with all human activities and a balance must be struck between the risks and benefits of those activities. Fortunately, virtually all activities of the upstream petroleum industry have effective technical options that can minimize or eliminate their environmental risks. Unfortunately, many of those options are expensive and may not be economically possible.

One of the keys to producing oil in environmentally responsible ways is to be aware of any potential hazards and to plan effective ways to minimize those hazards before a particular project begins. The first step in this process is education. Petroleum engineers, geologists, and managers must understand the place their industry occupies in society. All companies, including oil companies, exist by the grace and will of the people in society. If society does not want an industry to exist, that industry can be shut down, either through legislation, litigation, or economic boycotts. Unfortunately, the social pressures imposed on an industry are not necessarily based on accurate scientific information. Many existing regulations are politically based and do little to protect human health and the environment, yet they add considerable costs to businesses that must comply.

The environmental movement that has arisen over the past few decades has resulted in regulations that have had a profound effect on the operations of the upstream petroleum industry. These regulations have been imposed because the public no longer believes that the industry can regulate itself and still protect the environment. Some of this loss of confidence has been earned, but some is the result of deliberate misinformation spread by environmental extremists and a media willing to misrepresent the truth to sell copy.

Regardless of why the public lacks confidence in the ability of the petroleum industry to operate in an environmentally responsible manner, the industry must adapt and learn to live within the increasingly tight environmental regulations in order to survive. The fundamental shift in attitude toward proactive environmental protection that has begun must continue—the past ways of doing business are gone and will not return. It is not enough just to comply with whatever the current regulations might be; there must be a serious commitment toward protecting the environment in all activities, regardless of the regulations.

The key to effective regulations that protect the environment is for the regulations to be based on accurate scientific information. If an industry has lost its credibility with the public regarding environmental concerns due to its past behavior, then any accurate scientific information about the environmental impact of its current operations will also lack credibility. This results in regulations that are very costly to the industry, but do little to protect the environment.

Because funds available for environmental compliance are limited to those received within a project's minimum profitability level, these funds should be spent in ways that provide maximum protection for the environment. Bad regulations can require that available funds be spent in ways that provide little environmental protection. This increases the cost of doing business and can make many marginal projects uneconomical, resulting in a loss of jobs and reduction in domestic production. Thus, the conflict between the benefits of inexpensive energy and environmental protection are magnified by bad regulations.

The following hypothetical situation illustrates how misinformation and misunderstanding about sound scientific environmental principles can lead to the economic destruction of an industry:

A company applied for a discharge permit for a process and reported that the effluent concentrations of a particular chemical would be 75 parts per thousand. The discharge permit was denied on the grounds that the effluent concentration was too high. The company then spent thousands of dollars to upgrade their waste treatment stream and reduced the effluent concentration to 75 parts per million. Their discharge permit was again denied on the same grounds. The company then spent millions of dollars more to install the best available technology for treating the waste effluent. They successfully reduced the discharge concentration to 75 parts per billion. Unfortunately, the discharge permit was again denied on the grounds that the effluent concentration was still to high. The company then invested billions of dollars in research and development to create a new way to treat the effluent and lower the discharge concentration to 75 parts per trillion. The discharge permit was again denied. At this point, the company went bankrupt and was forced out of business because it spent all of its money trying to comply with environmental regulations. When they asked the permitting agency why their discharge permits were denied, they were simply told that 75 parts was just too high.

Although this story incorrectly implies that regulatory agencies do not base their regulations on sound scientific principles, the sad truth is that regulatory agencies must operate within laws passed by people who may lack an understanding of scientific environmental principles.

One industry that has been effectively destroyed by social pressure resulting from environmental misinformation is the nuclear power industry in the United States, even though the actual risks from nuclear power can be significantly lower than those from other, more acceptable forms of electrical power, such as coal. If the domestic petroleum industry completely loses the confidence of the public, it too can be effectively destroyed. If this occurs, then the imports of crude oil and products will increase significantly. Ironically, the transportation of imported crude oil creates a much greater environmental hazard than domestic production.

Historically, the petroleum industry has reacted often to new regulations by changing operational practices the minimum amount required to meet the letter of the regulations. But because of the complex, rapidly changing regulatory environment, this approach can no longer be used productively. Activities that comply completely with today's regulations can result in significant liability tomorrow.

Perhaps the most important thing the petroleum industry can do is adopt an attitude of working in harmony with the public will. Regulatory agencies should not be viewed as enemies but as co-workers in an effort to produce oil in both economically and environmentally sound ways. Conversely, regulatory agencies can do their part by imposing regulations based on accurate scientific information, not the prevailing political pressures. Mutual education between regulators, the petroleum industry, and the public at all levels is an important step in environmentally-responsible, cost-effective operations.

This partnership requires cooperation, teamwork, commitment, credibility, and trust among all parties involved in the exploration for and production of oil, including operating company managers, engineers, geologists, contractors, subcontractors, work crews, regulators, courts, and legislators. Environmentally related activities must be oriented toward improved environmental awareness and protection, not the avoidance of responsibility for environmental protection. Environmental awareness must be an integral part of everyone's daily job.

This type of attitude toward environmental responsibility has been formally adopted as a set of principles by the American Petroleum Institute member companies. These principles are known as the *Guiding Principles for Environmentally Responsible Petroleum Operations.*

Guiding Principles for Environmentally
Responsible Petroleum Operations

Recognize and respond to community concerns about raw materials, products, and operations.

Operate plants and facilities and handle raw materials and products in a manner that protects the environment and the safety and health of employees and the public.

Make safety, health, and environmental considerations a priority in planning and development of new products and processes.

Advise promptly appropriate officials, employees, customers, and the public of information of significant industry related safety, health, and environmental hazards and recommend protective measures.

Counsel customers, transporters, and others in the safe use, transportation, and disposal of raw materials, products, and waste materials.

Economically develop and produce natural resources and conserve those resources by using energy efficiency.

Extend knowledge of conducting or supporting research on the safety, health, and environmental effects of raw materials, products, processes, and waste materials.

Reduce overall emissions and waste generation.

Work with others to resolve problems created in disposal of hazardous substances from operations.

Participate with government and others in creating responsible laws, regulations, and standards to safeguard the community, workplace, and environment.

Promote these principles and practices by sharing experiences and offering assistance to others who produce, handle, use, transport, or dispose of similar raw materials, petroleum products, and wastes.

Source: American Petroleum Institute, 1992. Reprinted by permission of the American Petroleum Institute.

The benefits of being proactive in protecting the environment, as opposed to simply reacting to legislative, regulatory, or court-ordered mandates, can actually lower the long-term costs of doing business. For example, voluntary waste reduction and site remediation activities could result in the cleanup of a site at costs up to six times lower than if a regulatory agency mandates the cleanup, even if the identical remediation methods and standards are used (Knowles, 1992).

REFERENCES

American Petroleum Institute, "Oil and Gas Industry Exploration and Production Wastes," API Publication 471-01-09, Washington, D.C., July 1987.

American Petroleum Institute, "API Environmental Guidance Document: Onshore Solid Waste Management in Exploration and Production Operations," Washington, D.C., Jan. 1989.

American Petroleum Institute, "RP9000, Management Practices: Self-Assessment Process, and Resource Materials," Washington, D.C., Dec. 1992.

American Petroleum Institute, *Basic Petroleum Data Handbook,* Vol. 13, No. 3, Washington, D.C., Sept. 1993.

Bleier, R., Leuterman, A. J. J., and Stark, C., "Drilling Fluids Making Peace with the Environment," *J. Pet. Tech.,* Jan. 1993, pp. 6–10.

Derkies, D. L. and Souders, S. H., "Pollution Prevention and Waste Minimization Opportunities for Exploration and Production Operations," paper SPE 25934 presented at the Society of Petroleum Engineers/Environmental Protection Agency's Exploration and Production Environmental Conference, San Antonio, TX, March 7–10, 1993.

Deuel, L. E., "Evaluation of Limiting Constituents Suggested for Land Disposal of Exploration and Production Wastes," Proceedings of the U.S. Environmental Protection Agency's First International Symposium on Oil and Gas Exploration and Production Waste Management Practices, New Orleans, LA, Sept. 10–13, 1990, pp. 411–430.

Gilliland, A., *Environmental Reference Manual for the Oil and Gas Exploration and Producing Industry,* Texas Independent Producers and Royalty Owners Association, Austin, TX, 1993.

Godec, M. L. and Biglarbigi, K., "Economic Effects of Environmental Regulations of Finding and Developing Crude Oil in the U.S.," *J. Pet. Tech.,* Jan. 1991, pp. 72–79.

Gray, P. R., "NORM Contamination in the Petroleum Industry," *J. Pet. Tech.,* Jan. 1993, pp. 12–16.

Guckian, W. M., Hurst, K. G., Kerns, B. K., Moore, D. W., Siblo, J. T., and Thompson, R. D., "Initiating an Audit Program: A Case History," paper SPE 25955 presented at the Society of Petroleum Engineers/Environmental Protection Agency's Exploration and Production Environmental Conference, San Antonio, TX, March 7–10, 1993.

Hudgins, C. M., Jr., "Chemical Treatments and Usage in Offshore Oil and Gas Production Systems," *J. Pet. Tech.,* May 1992, pp. 604–611.

Interstate Oil Compact Commission, *EPA/IOCC Study of State Regulation of Oil and Gas Exploration and Production Waste,* Interstate Oil Compact Commission, Oklahoma City, OK, Dec. 1990.

Knowles, C. R., "A Responsible Remediation Strategy," Proceedings of Petro-Safe '92, Houston, TX, 1992.

Mount, D. R., Gulley, D. D., and Evans, J. M., "Salinity/Toxicity Relationships to Predict the Acute Toxicity of Produced Waters to Freshwater Organisms," paper SPE 26007 presented at the Society of Petroleum Engineers/Environmental Protection Agency's Exploration and Production Environmental Conference, San Antonio, TX, March 7–10, 1993.

National Research Council, *Oil in the Sea: Inputs, Fates, and Effects.* Washington, D.C.: National Academy Press, 1985.

Perkins, J., "Cost to Petroleum Industry of Major New and Future Federal Government Environmental Regulations," American Petroleum Institute, Discussion Paper #070, Oct. 1991.

Perry, C. W. and Gigliello, K., "EPA Perspective on Current RCRA Enforcement Trends and Their Application to Oil and Gas Production Wastes," Proceedings of the U.S. Environmental Protection Agency's First International Symposium on Oil and Gas Exploration and Production Waste Management Practices, New Orleans, LA, Sept. 10–13, 1990, pp. 307–318.

Savage, L. L., "Even If You're On the Right Track, You'll Get Run Over If You Just Sit There: Source Reduction and Recycling in the Oil Field," paper SPE 26009 presented at the Society of Petroleum Engineers/Environmental Protection Agency's Exploration and Production Environmental Conference, San Antonio, TX, March 7–10, 1993.

Snavely, E. S., "Radionuclides in Produced Water," report prepared for the API Guidelines Steering Committee, American Petroleum Institute, Washington, D.C., 1989.

Stephenson, M. T., "Components of Produced Water: A Compilation of Industry Studies," *J. Pet. Tech.,* May 1992, pp. 548–603.

Stilwell, C. T., "Area Waste-Management Plans for Drilling and Production Operations," *J. Pet. Tech.,* Jan. 1991, pp. 67–71.

Sullivan, M. J., "Evaluation of Environmental and Human Risk from Crude-Oil Contamination," *J. Pet. Tech.,* Jan. 1991, pp. 14–16.

Thurber, N. E., "Waste Minimization for Land-Based Drilling Operations," *J. Pet. Tech.,* May 1992, pp. 542–547.

Wojtanowicz, A. K., Field, S. D., and Osterman, M. C., "Comparison Study of Solid/Liquid Separation Techniques for Oilfield Pit Closures," *J. Pet. Tech.,* July 1987, pp. 845–856.

CHAPTER 2

Drilling and Production Operations

In the upstream petroleum industry, there are two major operations that can potentially impact the environment: drilling and production. Both operations generate a significant volume of wastes. Environmentally responsible actions require an understanding of these wastes and how they are generated. From this understanding, improved operations that minimize or eliminate any adverse environmental impacts can be developed.

Drilling is the process in which a hole is made in the ground to allow subsurface hydrocarbons to flow to the surface. The wastes generated during drilling are the rock removed to make the hole (as cuttings), the fluid used to lift the cuttings, and various materials added to the fluid to change its properties to make it more suitable for use and to condition the hole.

Production is the process by which hydrocarbons flow to the surface to be treated and used. Water is often produced with hydrocarbons and contains a variety of contaminants. These contaminants include dissolved and suspended hydrocarbons and other organic materials, as well as dissolved and suspended solids. A variety of chemicals are also used during production to ensure efficient operations.

During both drilling and production activities, a variety of air pollutants are emitted. The primary source of air pollutants are the emissions from internal combustion engines, with lesser amounts from other operations, fugitive emissions, and site remediation activities.

2.1 DRILLING

The process of drilling oil and gas wells generates a variety of different types of wastes. Some of these wastes are natural byproducts of drilling through the earth, e.g., drill cuttings, and some come from materials used to drill the well, e.g., drilling fluid and its associated additives. This section reviews the drilling process, the drilling fluid composition, methods to separate cuttings from the drilling fluid, the use of reserves pits, and site preparation.

2.1.1　Overview of the Drilling Process

Most oil and gas wells are drilled by pushing a drill bit against the rock and rotating it until the rock wears away. A drilling rig and system is designed to control how the drill bit pushes against the rock, how the resulting cuttings are removed from the well by the drilling fluid, and how the cuttings are then removed from the drilling fluid so the fluid can be reused.

The major way in which drilling activities can impact the environment is through the drill cuttings and the drill fluid used to lift the cuttings from the well. Secondary impacts can occur due to air emissions from the internal combustion engines used to power the drilling rig.

During drilling, fluid is injected down the drill string and though small holes in the drill bit. The drill bit and holes are designed to allow the fluid to clean the cuttings away from the bit. The fluid, with suspended cuttings, then flows back to the surface in the annulus between the drill string and formation. At the surface, the cuttings are separated from the fluid; the cuttings, with some retained fluid, are then placed in pits for later treatment and disposal. The separated fluid is then reinjected down the drill string to lift more cuttings.

The base fluid most commonly used in the drilling process is water, followed by oil, air, natural gas, and foam. When a liquid is used as the base fluid, either oil-based or water-based, it is called "mud." Water-based drilling fluids are used in about 85% of the wells drilled worldwide. Oil-based fluids are used for virtually all of the remaining wells.

During the drilling process, some mud can be lost to permeable underground formations. To ensure that mud is always available to keep the well full, extra mud is always mixed at the surface and kept in reserves or mud pits for immediate use. Reserves pits vary in size, depending on the depth of the well. The pits can be up to an acre in area and be 5–10 feet deep. Steel tanks are also used for mud pits, especially in offshore operations. Pits are also used to store supplies of water, waste fluids, formation cuttings, rigwash, and rainwater runoff.

2.1.2 Drilling Fluids

Drilling fluids serve a number of purposes in drilling a well. In most cases, however, the base fluid does not have the proper physical or chemical properties to fulfull those purposes, and additives are required to alter its properties. The primary purpose of drilling fluid is to remove the cuttings from the hole as they are generated by the bit and carry them to the surface. Because solids are more dense than the fluid, they will tend to settle downward as they are carried up the annulus. Additives to increase the fluid viscosity are commonly used to lower the settling velocity.

Drilling fluids also help control the well and prevent blowouts. Blowouts occur when the fluid pressure in the wellbore is lower than the fluid pressure in the formation. Fluid in the formation then flows into the wellbore and up to the surface. If surface facilities are unable to handle this flow, uncontrolled production can occur. The primary fluid property required to control the well is the fluid's density. Additives to increase fluid density are commonly used.

Drilling fluids also keep the newly drilled well from collapsing before steel casing can be installed and cemented in the hole. The pressure of the fluid against the side of the formation inhibits the walls of the formation from caving in and filling the hole. Additives are often used to prevent the formation from reacting with the base fluid. One common type of reaction is shale swelling.

A final function of drilling fluids is to cool and lubricate the drill bit as it cuts the rock and lubricate the drill string as it spins against the formation. This extends the life of the drill bit and reduces the torque required at the rotary table to rotate the bit. Additives to increase the lubricity of the drilling fluid are commonly used, particularly in highly deviated or horizontal wells.

Many of the additives used in drilling fluids can be toxic and are now regulated. To comply with new regulations, many new additives have been formulated (Clark, 1994). These new additives have a lower toxicity than those traditionally used, thus lowering the potential for environmental impact.

Water-based Drilling Fluids

Water is the most commonly used base for drilling fluids or muds. Because it does not have the physical and chemical properties needed to fulfill all of the requirements of a drilling mud, a number of additives are used to alter its properties. During drilling, formation materials get incorporated into the drilling fluid, further altering its composition and properties. A typical elemental composition of common constituents of water-based drilling muds is given in Table 2-1 (Deeley, 1990). These constituents are discussed in more detail below.

Viscosity Control

One of the most important functions of a drilling fluid is to lift cuttings from the bottom of the well to the surface where they can be removed. Because cuttings are more dense than water, they will settle downward through the water from gravitational forces. The settling velocity is controlled primarily by the viscosity of the water and the size of the cuttings. Because the viscosity of water is relatively low, the settling velocity for most cuttings is high. To remove the cuttings from the well using water only, a very high water velocity would be required. To lower the settling velocity of cuttings and decrease the corresponding mud circulation rate, viscosifiers are added to the water to increase its viscosity.

The most commonly used viscosifier is a hydratable clay. Some clays, like smectite, consist of molecular sheets with loosely held cations between them, such as Na^+. If the clay is contacted with water having a cation concentration that is lower than the equilibrium concentration for the cation in the clay, the cation atom between the sheets can be exchanged with water molecules. Because water molecules are physically larger than most cations, the spacing between the clay sheets expands and the clay swells (hydrates). During the mixing and shearing that occurs as water is circulated through the well, these clay sheets can separate, forming a suspension of very small solid particles in the water. The viscosity of this suspension is significantly higher than that of pure water and is more effective in lifting the larger formation cuttings out of the well.

The most common clay used is Wyoming bentonite. This clay is composed mostly of sodium montmorillonite, a variety of smectite.

Table 2-1
Elemental Composition of Drilling Fluid Constituents (mg/kg)

Element	Water	Cuttings	Barite	Clay	Chrome-lignosulfonate	Lignite	Caustic
Aluminum	0.3	40,400	40,400	88,600	6,700	6,700	0.013
Arsenic	0.0005	3.9	34	3.9	10.1	10.1	0.039
Barium	0.01	158	590,000	640	230	230	0.26
Calcium	15	240,000	7,900	4,700	16,100	16,100	5,400
Cadmium	0.0001	0.08	6	0.5	0.2	0.2	0.0013
Chromium	0.001	183	183	8.02	40,030	65.3	0.00066
Cobalt	0.0002	2.9	3.8	2.9	5	5	0.00053
Copper	0.003	22	49	8.18	22.9	22.9	0.039
Iron	0.5	21,900	12,950	37,500	7,220	7,220	0.04
Lead	0.003	37	685	27.1	5.4	5.4	0.004
Magnesium	4	23,300	3,900	69,800	5,040	5,040	17,800
Mercury	0.0001	0.12	4.1	0.12	0.2	0.2	5
Nickel	0.0005	15	3	15	11.6	11.6	0.09
Potassium	2.2	13,500	660	2,400	3,000	460	51,400
Silicon	7	206,000	70,200	271,000	2,390	2,390	339
Sodium	6	3,040	3,040	11,000	71,000	2,400	500,000
Strontium	0.07	312	540	60.5	1030	1030	105

Source: Deeley, 1990.

Most drilling fluids are composed of 3% to 7% bentonite by volume. Other clays can be used, but typically do not provide as high a mud viscosity for the same amount of clay added. During normal drilling operations, natural clays in the formations can also be incorporated into the mud, increasing the clay content and mud viscosity over time.

Adding hydratable clays to the water used as a drilling fluid provides a second important benefit for drilling of wells. Because the pressure of the mud in the wellbore is normally kept above the pressure in the formation to prevent blowouts, the water (mud filtrate) will flow into a permeable formation and be lost. When this occurs, the suspended clays are filtered out at the face of the formation, building a mudcake along the walls of the well. The clay particles of this mudcake are virtually always smaller than the grains of a permeable formation, so the resulting permeability of the mudcake is much lower than that of the formation. This low permeability mudcake acts as a barrier to minimize subsequent fluid losses to the formation. Because fluid losses are lower, the total volume of mud needed to drill the well is reduced.

One difficulty with using clay particles for viscosity control is that they tend to flocculate (agglomerate) if the mud is allowed to remain static in the wellbore. When flocculation occurs, the mud viscosity can significantly increase. If the viscosity becomes too high, the mud can become too difficult to pump at reasonable pressures and flow rates, rendering it ineffective as a drilling fluid. Flocculation occurs when the electrostatic charges along the periphery of the clay particles are allowed to attract other clay particles. The flocculation rate increases with an increasing clay content and electrolyte (salt) concentration in the mud.

A variety of materials are available that can suppress flocculation of clay particles in drilling muds, although none are totally effective under all conditions. The most common deflocculants are phosphates, tannins, lignites, and lignosulfonates. Phosphate deflocculants can be used when the salt concentrations and temperatures are low. Tannins are effective in moderate concentrations of electrolyte concentration and moderate temperatures. Lignites and lignosulfonates can be effective at high temperatures, particularly if they are complexed with heavy metals like chromium.

Polymers, like xanthan gum, have also been developed to increase the viscosity of drilling mud. These polymers have the advantage of shear thinning, which lowers the viscosity and required pumping power during high pumping rates, when a high viscosity is not needed.

Density Control

Another important function of a drilling fluid is to control the fluid pressure in the wellbore. Because many formations are hydrostatically pressured or overpressured and the pressure in the wellbore must be kept higher than that in the formation, the pressure in the wellbore must normally be higher than the hydrostatic pressure for pure water to prevent the well from blowing out. The fluid pressure in the wellbore is controlled by varying the density of the drilling fluid. The density is varied by adding heavy solids to the fluid.

Although the clays added to control the fluid viscosity also increase the fluid density, their specific gravity of 2.6 and low concentration in the mud is insufficient to provide the needed density for many applications. Materials having a higher specific gravity are normally required to obtain the desired mud density.

The most common material used to increase the density of drilling mud is barite (barium sulfate, $BaSO_4$). Barite has a high specific gravity of 4.2. In some wells requiring a very high density, barite can constitute as much as 35% of the drilling fluid by volume. Because of the high specific gravity of barite, viscosity control additives (clays) are normally used to keep the barite suspended in the fluid.

Other materials that can be used to control drilling fluid density include calcium carbonate, iron carbonate, ilmenite ($FeO\text{-}TiO_2$) and hematite (Fe_2O_3). These materials are harder than barite and are less susceptible to particle size reduction during drilling. Although these materials have a lower specific gravity than barite, they have the added benefit of lowering the barium concentration in the drilling mud. Galena (PbS) can also be used, but will result in lead being added to the drilling mud. Rarely, barium carbonate has been used.

Lost Circulation Control

During drilling, fluid is lost to the formation as drilling fluid leaks into permeable strata. To minimize this loss, small particles are added to drilling fluids that will filter out on the formation face as fluid is lost. These solids then form a low permeability mudcake that limits further fluid loss. In most cases, the clay particles added to control the viscosity of a drilling fluid are successful in controlling fluid loss to the formation.

In some formations, however, the pore sizes may be so large that the clay particles are unable to bridge the pores and build a filter cake. Such formations may include those having natural or induced fractures, very high permeability sands, or vugs. To limit fluid loss in such formations, larger solids can be added to the drilling fluid. A mudcake of clay particles is then built on the bridge created by those solids. Solids that are commonly used for this application include mica, cane fibers, ground nutshells, plastic, sulfur, perlite, cellophane, cottonseed hulls, and sawdust.

If solids cannot be used to build a filter cake, the viscosity of the drilling fluid can be increased to limit fluid loss. Water-soluble polymers like starch, sodium polyacrylate, and sodium carboxymethylcellulose can be used.

pH Control

A high mud pH between 9.5 and 10.5 is almost always desired in drilling operations. A high pH suppresses the corrosion rate of drilling equipment, minimizes hydrogen embrittlement of steel if hydrogen sulfide enters the mud, lowers the solubility of calcium and magnesium to minimize their dissolution, and increases the solubility of ligno-sulfonate and lignite additives. A high pH is also beneficial for many new organic viscosity control additives. To keep the pH in the desired range, caustic (sodium hydroxide) is normally added to the mud. Some of the new polymer muds, however, have better shale stabilization properties at a lower pH (Clark, 1994).

Lubricants

During drilling, a considerable amount of friction can be generated between the drill bit and formation and between the drill string and wellbore walls, particularly for deviated and horizontal wells. To reduce this friction, lubricants are sometimes added to drilling fluids. These lubricants speed drilling and help maintain the integrity of the well. Common lubricants include diesel oil, mineral/vegetable oils, glass beads, plastic beads, wool grease, graphite, esters, and glycerols.

If a drill string becomes stuck in a well, a lubricant is usually circulated through the well to help free it. These *spotting fluids* have traditionally been formulated with diesel or mineral oils. Because these fluids "contaminate" cuttings with a hydrocarbon, the discharge and disposal options for cuttings is limited in some areas. Water-based spotting fluids are also available (Clark and Almquist, 1992).

Corrosion Inhibitors

Corrosion is commonly caused by dissolved gases in the drilling mud, e.g., oxygen, carbon dioxide, or hydrogen sulfide. Optimum corrosion protection of drilling equipment would include elimination of these gases from the mud. If elimination is not possible, the corrosion rate should be reduced. A wide variety of chemicals are available to inhibit corrosion from drilling mud. These additives are often used even when the pH is maintained in the desired range.

Corrosion inhibitors do not prevent corrosion, but reduce the corrosion rate to acceptable levels, e.g., below 400 mills per year or 0.02 lbm metal per ft^2 of metal in 10 hours. Inhibitors coat the metal surface and limit the diffusion rate of corrosive chemicals to the surface. The most common inhibitors utilize a surfactant that protects the metal with a coating of oil. High molecular weight morpholines and filming amines are most commonly used for oilfield applications. Ethylene diamine tetracetic acid (EDTA) is sometimes used to dissolve pipe corrosion.

Oil-soluble organic inhibitors applied every 10 hours appear to successfully reduce oxygen corrosion. These inhibitors are strongly absorbed on clays and cuttings, however, increasing the amount of inhibitor required. Water-soluble organic corrosion inhibitors may not be effective for controlling oxygen corrosion, although they can be used to reduce pitting from H_2S in the absence of oxygen. A more complete discussion of corrosion is given by Jones (1988).

Biocides

Sulfur reducing bacteria can grow in many drilling muds, particularly those containing starches and polymer additives. These bacteria can degrade the mud and can enter the formation, where they can sour the reservoir (generate hydrogen sulfide gas). Hydrogen sulfide causes corrosion of equipment when present in drilling muds. To prevent these bacteria from growing, biocides are added to drilling fluids. Common biocides include paraformaldehyde, chlorinated phenol, isothiazolin, and glutaraldehyde. The latter two biocides have lower toxicities and are replacing the former two in popularity (Clark, 1994).

Formation Damage Control

Many formations contain active clays that swell upon contact with fresh water. These swelling clays can plug pores in the reservoir, lowering its permeability, or they can cause shale around the wellbore to slough into the wellbore, "wellbore washout." To prevent these reactions from occurring, salts are commonly added to the drilling fluid. These salts prevent water molecules from exchanging with the cations in the clays. Salts commonly used include sodium and potassium chloride. Potassium acetate or potassium carbonate can also be used, as well as cationic polymers. Shale stabilization additives based on glycols have also been successfully used (Reid et al., 1993). A number of cationic polymer muds having good shale stabilization properties have also been introduced (Clark, 1994).

A related problem during drilling is that cuttings can ball around the bit, forming a gummy paste. This paste reduces drilling speed because it is not easily removed from the bit by the drilling fluid. Copolymer/polyglycol muds have been successfully used to prevent bit-balling (Enright and Smith, 1991).

If a well is drilled through a salt dome, a water-based mud that is saturated in chloride salts may be required to prevent excessive dissolution of the salt along the wellbore.

Oil-based Drilling Fluids

Various organic fluids are also used as a base for drilling muds. In some cases, the properties of these "oil-based" muds are superior to those of water-based muds. Like water, however, these organic fluids do not have all of the proper physical and chemical properties needed to fulfill all of the requirements of a drilling mud, so various additives are also used.

Oil-based muds are often preferred for high-temperature wells, i.e., wells with temperatures greater than about 300°F. At temperatures above that level, many of the additives used with a water-based fluid can break down.

Oil-based muds are also used in wells containing water-sensitive minerals, e.g., salt, anhydrite, potash, gypsum, or hydratable clays and shales. Using an oil-based mud in a reactive formation can reduce wellbore washout by more than 20% (Thurber, 1990). Reducing the amount of washout reduces both the volume of drill cuttings to be disposed of and the volume of drilling fluid required to drill the hole. Reducing interactions between the drilling fluid and formation minerals by using an oil-based mud also limits the degradation of cuttings into smaller particles, which improves the efficiency of separating the solids from the drilling fluid.

Oil-based muds are also used in wells containing reactive gases like CO_2 or H_2S. When oil-based muds are used, corrosion is minimized because the continuous oil phase does not act as an electrolyte. These gases are prime contributors to corrosion of drilling equipment in water-based mud systems.

Another application of oil-based muds is in wells requiring unusually high levels of lubrication between the drill pipe and the formation. These wells include deviated or horizontal wells, where the drill pipe rotates against the formation over long intervals. Oil-based muds are also useful for freeing pipe that has become stuck in the well.

Oil-based muds are generally more expensive than water-based muds and have a greater potential for adverse environmental impact. The benefits of oil-based muds, however, can result in a significant savings in the cost of drilling a well. Because of their superior properties, drilling can often be completed faster, which may result in lower overall environmental consequences than those of water-based muds. Because oil-based muds are more expensive, they are also more likely to be reconditioned and reused than water-based muds.

Historically, the most common base oil used has been #2 diesel. It has an acceptable viscosity, low flammability, and a low solvency for any rubber in the drilling system. Diesel, however, is relatively toxic, making the environmental impact of diesel-based muds generally higher than those of water-based muds.

The most common additive used in oil-based muds for viscosity control is water in the form of a water-in-oil emulsion. Small, dispersed drops of water in the continuous oil phase can significantly increase the mud viscosity. Water contents of typically 10% have been used. A chemical emulsifier (surfactant) is normally added to prevent the water droplets from coalescing and settling from gravitational forces. Commonly used emulsifiers are calcium or magnesium fatty-acid soaps. If further viscosity increases are required, solids can be added to the mud, including asphalts, amine-treated bentonite, calcium carbonate, or barite.

The density of oil is significantly lower than that of water, so density control additives normally must be used. The water in water-in-oil emulsions only slightly increases the mud density, so solids are normally added. The same solids that are used to increase the viscosity—asphalts, amine-treated bentonite, calcium carbonate, or barite—can be used to increase the density. One limitation with oil-based muds is that most of the solids that enter the mud, including cuttings, are water-wet. To prevent the solids from concentrating in the dispersed water droplets and settling out, chemical wettability agents (surfactants) are added to change the wettability of the solids to oil-wet. This allows the solids to be dispersed through the more voluminous oil phase.

One of the advantages of oil-based muds is their compatibility with water-sensitive formations. Because the continuous phase is oil, only oil can enter the formation as a filtrate. Water invasion is severely limited, which minimizes the damage to the formation. Because clay particles do not flocculate in oil-based muds, bit-balling is also minimized. If fluid loss becomes too high, fluid loss agents like bentonite, asphalt, polymers, manganese oxide, and amine-treated lignite can be used.

Although oil-based muds have a lower corrosion rate than water-based muds, corrosion can occur, particularly when drilling through a formation containing CO_2 or H_2S. Like water-based muds, the primary method to control corrosion is to control the pH of the water phase of the mud. A common additive for pH control of oil-based muds is calcium oxide.

A number of oil-based muds using organic materials have been developed as low-toxicity substitutes for diesel (Friedheim and Shinnie, 1991; Peresich et al. 1991). Mineral and synthetic oils are becoming increasingly popular as a base for drilling mud (Clark, 1994).

Unwanted Components

All drilling muds generally have a number of unwanted components that can potentially harm the environment. The most common of these are heavy metals, salt, and hydrocarbons. The concentration of these materials varies significantly. The primary concern arises when the drilling fluid must be disposed of.

Heavy Metals

Heavy metals can enter drilling fluids in two ways: Many metals are naturally occurring in most formations and will be incorporated into the fluid during drilling; other metals are added to the drilling fluid as part of the additives used to alter the fluid properties. The most commonly found metals have traditionally been barium from barite weighting agents and chromium from chrome-lignosulfonate deflocculants.

Heavy metals naturally occur in most rocks and soils, although at relatively low concentrations. The elemental concentrations of native soils and gravels on the Alaskan North Slope are summarized in Table 2-2. Although the concentrations of the major elements will vary from carbonate to siliceous rocks, the concentration of the trace elements, including heavy metals, is probably representative of rocks and soils of many other areas. Naturally occurring metals of particular concern include arsenic, barium, cadmium, chromium, lead, and mercury.

Drilling fluids typically contain high concentrations of barium. Barium is a constituent of barite, which is used as a density control material. The most commonly used form of barium, however, is barium sulfate, which is highly insoluble. Because of its low solubility, it will not leach with groundwater movement, nor will it be taken up by plants and enter the food chain.

Chromium is another major constituent of many mud additives, particularly chrome-based deflocculants. Chromium in its toxic hexavalent form can be used as a gel inhibitor/thinner, a high-temperature stabilizer, a dispersant, a biocide, and a corrosion inhibitor. It is believed, however, that hexavalent chromium is quickly reduced to its relatively nontoxic trivalent form in a mud system (Campbell and Akers, 1990). Typical chromium levels in drilling muds are between 100 and 1,000 mg/L (Bleier, Leuterman, and Stark, 1993).

Another significant source of heavy metals in drilling fluid is the thread compound (pipe dope) used on the pipe threads when making up a drill string. Pipe dope serves two primary purposes: (1) it prevents the seizure of the joint from galling at high stresses and (2) it seals the joint and prevents fluid flow along the threads. Early formulations

Table 2-2
Composition of Alaskan North Slope Soils and Gravels

Element	Mean Level (mg/kg dry)	Standard Deviation
Aluminum	7,050	6,180
Arsenic	1.83	2.15
Barium	397	802
Boron	29.4	27.4
Cadmium	0.153	0.185
Calcium	46,700	65,500
Chromium	11	16
Copper	14	11.9
Iron	19,600	15,800
Lead	4.24	5.06
Magnesium	3,440	3,880
Manganese	484	1,040
Mercury	0.268	0.289
Nickel	21.3	14.8
Potassium	699	810
Selenium	0.267	0.187
Silicon	1,640	6,230
Silver	0.26	0.177
Sodium	529	672
Strontium	93.4	140
Vanadium	29.9	140
Zinc	74.6	46.1

Source: from Schumacher et al., 1991.
Copyright SPE, with permission.

of pipe dope contained as much as 60% metals by weight, primarily lead, zinc, copper, or combinations of these metals (McDonald, 1993). These metals are malleable and deform within the threads without fracturing, forming both a seal and lubricant for the threads. These metals, however, can leach out of the pipe dope and contaminate the drilling fluid, particularly if an excess of pipe dope is used.

Another source of heavy metals in drilling fluid is from crude oil. Crude oil naturally contains widely varying concentrations of various heavy metals. These metals can enter the drilling fluid during drilling through a formation containing crude oil or if a kick occurs and oil flows into the well. Metals found in crude oil include aluminum, boron, calcium, chromium, cobalt, copper, gold, iron, lead, magnesium, manganese, nickel, phosphorus, platinum, silicon, silver, sodium, strontium, tin, uranium, and vanadium. Of these elements, vanadium and nickel occur in the highest concentrations. The concentration of metals in some crude oils is typically on the order of a few parts per million to a few tens of parts per million, although concentrations as high as thousands of parts per million have been reported (National Research Council, 1985).

A number of other metals are found in drilling fluid additives, although at lower concentrations. Arsenic can be used as a biocide to prevent the growth of bacteria. Cadmium is found in some pipe dopes. The mineral barite, the source for the barium sulfate used for density control, can have relatively high naturally occurring levels of cadmium and mercury (Candler et al., 1990). Mercury has also been used in manometers in the natural gas industry to meter the flow rate of gas. Zinc is occasionally used as inorganic zinc salts for density control or as hydrogen sulfide scavengers to minimize corrosion and maintain human safety.

Salt

Another unwanted component of drilling fluid at disposal time are salts. Salts, like sodium or potassium chloride, are often added to drilling fluid to protect sensitive formations from reacting with the drilling fluid. The salt concentration of a drilling fluid can also significantly increase if a well is drilled through a salt dome or a formation having water with a high salt concentration.

Hydrocarbons

Except for oil-based muds, hydrocarbons are normally an undesirable material in drilling mud because they contaminate the cuttings. Hydrocarbons enter a mud while drilling through a hydrocarbon-bearing formation or when oil is used for a spotting fluid when a pipe becomes stuck. In general, the deeper the well, the greater the concentration of hydrocarbons that enter the mud.

2.1.3   Drilling Fluid Separations

During the drilling process, a large volume of cuttings are generated and carried out of the well by the drilling fluid. These cuttings must be separated from the mud liquid so the liquid can be reinjected into the drill string to remove more cuttings. Cuttings contaminated with drilling mud are a major source of petroleum industry waste. The potential environmental impact of such cuttings can be significantly reduced by separating the solid cuttings from the more toxic mud.

The effectiveness of separating cuttings from the mud depends primarily on the cuttings size. Separations can be enhanced if the cuttings size is kept as large as possible. Cuttings size depends on a number of factors. The most important factor in keeping cuttings size large is to generate large cuttings at the bit during drilling. The initial cuttings size is controlled by the bit type, the weight on bit, and the formation type. A second factor in controlling the cuttings size is to minimize additional grinding of the cuttings in the well as they are lifted to the surface. Cuttings removal is controlled by the hydraulic design of the bit jets, the mud viscosity, the mud velocity, the well depth, the rotational speed of the drill string, and the mechanical strength of the cuttings. A third factor controlling cuttings size is whether the cuttings contain clays which can hydrate (deflocculate) in the mud before separation. Clay hydration can be controlled by the mud chemistry. Additives like polyacrylamides, polymers and salts, as well as oil-based muds, can help control formation reactivity and minimize degradation of solids.

The first stage of separation is to remove large cuttings from the mud with a shale shaker. Shale shakers are vibrating screens over which the mud passes. The liquid and small cuttings pass through the screens, while the larger cuttings remain on the screen. If the mud contains gas, the shale shaker will also separate much of it from the mud. The mud and small cuttings that pass through the screens are returned to the mud pit, where additional separation of cuttings and gases occurs from gravitational settling. The effectiveness of vibrating screens depends on the vibrator placement, vibration frequency, vibration amplitude, speed of solids as they pass across the screens, and screen opening size (Hoberock, 1980; Lal and Hoberock, 1988).

Chemicals can be added to the mud that cause the small clay particles to coagulate or flocculate into larger groups of particles

(American Petroleum Institute, 1990b). The larger flocculates then settle more rapidly in the mud pits. This process involves the neutralization of the surface charge (zeta potential) on suspended particles to overcome coulombic electrical repulsion between the particles and allow aggregates to form. Inducing alternating electrical currents to overcome the coulombic repulsion has also been proposed (Farrell, 1991).

If a drilling mud contains gas that is not removed by the solids separation equipment, a vacuum chamber can be added to the mud system. This lowers the mud pressure in the chamber and expands the size of the gas bubbles, allowing them to be separated from the liquid by gravity more rapidly. In these systems, the mud is typically passed over inclined planes in thin layers to enhance separation.

If the proper equipment and procedures are not used to remove the cuttings as they are added to the mud system, the concentration of cuttings in the mud gradually increases with time, and the mud properties, such as density and viscosity, are degraded. The maximum tolerable solids concentration varies with the mud used, but is generally between 4% and 15% (Wojtanowicz, 1991). To maintain the mud properties in the desirable range, the mud can be diluted; this requires the addition of more base fluid, either water or oil, and many of the chemicals needed to alter its chemical properties. Dilution, however, increases the volume of drilling waste that must ultimately be disposed of.

In many cases, shale shakers and settling pits are insufficient to separate the mud solids from liquids, and further treatment with advanced technology is required. For example, after separating the solids from the mud, a significant volume of liquid is normally retained with the cuttings. Volumetric measurements from offshore platforms have shown that the total volume of liquids with the cuttings after discharge can be from 53% to 73% (Wojtanowicz, 1991). In some cases, further dewatering of the solids may be required before disposal. Advanced separation methods are discussed in Chapter 6.

One difficulty with using advanced technology for improved separations at a drill site is the high cost of equipment rental. The expenditure for this equipment can be easier to justify if a good economic model for their benefits is used. One such model has been proposed by Lal (1988) and was subsequently verified by field performance (Lal and Thurber, 1989).

2.1.4   Reserves Pits

The most common method for the disposal of drilling wastes for onshore wells is in on-site reserves pits. The contents of reserves pits vary, depending on the drilling mud and the types of formations drilled. Reserves pits, however, can cause local environmental impact, particularly older pits that contain materials that are currently banned from such disposal or that were not constructed according to current regulations. The environmental impact of modern reserves pits are minimal.

The composition of the fluid in a reserves pit may be different from that of the original drilling fluid. Chemical and physical alterations of drilling fluids can occur during and after drilling from the heat and pressure encountered during drilling or from the addition of formation materials. Other materials may also be added to the pit before closure, either deliberately or inadvertently. Such materials include caustic soda, rig wash, diesel fuel, waste oil from machinery, metal and plastic containers, and other refuse (Powter, 1990). Bad storage and disposal practices associated with reserves pits have lead to their being a source of benzene, lead, arsenic, and fluoride, even when these components were not detected in the active mud system (Wojtanowicz, 1991).

The heavy metals and other dissolved solids contents in both the water and mud (sludge) phases of 125 reserves pits scattered around the United States were measured in one study, and the total and water-soluble (leachable) concentrations were determined (Leuterman et al., 1988). The mean metals concentrations of all of the pits varied significantly with species, with mean concentrations on the order of a few tens of mg/L. These data are summarized in Table 2-3. It was found that the metals concentrations in the mud phase were generally higher than in the water phase, indicating that most of the metals were probably bound to the organic and clay particles and were not readily available for leaching.

In separate studies, the heavy metals contents of reserves pits in the U.S. Gulf Coast were also analyzed and found to vary significantly (Wojtanowicz et al., 1989 and Deuel and Holliday, 1990). In the latter study, the pit contents were analyzed by the U.S. Environmental Protection Agency, the American Petroleum Institute, and in a private study under Louisiana State guidelines. The analysis protocols and procedures differed in the three studies and yielded somewhat different results. The results are summarized in Tables 2-4 and 2-5.

Table 2-3
Average Elemental Composition of Reserves Pits

Metal	Phase	Concentration (mg/L)
Calcium	Mud	207
	Water	156
Chromium (soluble)	Mud	3.97
	Water	2.09
Chromium (total)	Mud	56.05
	Water	14.47
Lead (soluble)	Mud	6.51
	Water	0.08
Lead (total)	Mud	24.46
	Water	3.36
Magnesium (total)	Mud	17.21
	Water	65.47
Manganese (soluble)	Mud	0.29
	Water	0.19
Manganese (total)	Mud	77.67
	Water	4.74
Potassium	Mud	313
	Water	750
Sodium	Mud	1,819
	Water	2,125
Zinc (soluble)	Mud	0.21
	Water	0.07
Zinc (total)	Mud	52.54
	Water	5.07
pH	Mud	8.79
	Water	8.10
Carbonate	Mud	135
	Water	56
Chloride	Mud	2,204
	Water	3,639
Bicarbonate	Mud	582
	Water	447
Hydroxyl	Mud	45
	Water	0.47
Sulfate	Mud	929
	Water	551

Source: from Leuterman et al., 1988.
Copyright SPE, with permission.

Table 2-4
Average Elemental Composition of Reserves Pits

Metal	Pit 1 (mg/g)	Pit 2 (mg/g)	Pit 3 (mg/g)	Pit 4 (mg/g)
Barium	10.119	8.906	11.088	7.085
Chromium	0.071	0.024	0.179	0.056
Lead	0.044	0.354	0.057	0.037
Zinc	0.170	0.256	0.148	0.162

Source: from Wojtanowicz et al., 1989.
Copyright SPE, with permission.

Table 2-5
Average Elemental Composition of Reserves Pits

Metal	Private Study (mg/g)	API (mg/g)	EPA (mg/g)
Arsenic	0.003	0.008	0.029
Calcium	31.0	47.2	71.7
Chromium	0.016	0.017	0.081
Barium (total)	29.2	N/A	N/A
Iron	15.1	21.2	56.8
Lead	0.064	0.059	0.446
Magnesium	3.72	4.72	8.10
Manganese	0.273	0.393	0.940
Potassium	2.61	1.85	N/A
Sodium	2.36	3.78	5.62
Zinc	0.120	0.189	0.683

Source: from Deuel and Holliday, 1990.
Copyright SPE, with permission.

The heavy metals found in pits are not uniformly distributed in the pits. Heavy metals are often bound to coarse particulates and tend to accumulate near the point of discharge. The nonuniform distribution of metals in a pit needs to be considered when sampling the pit for metals concentration (Deuel and Holliday, 1990). Other studies, however, reveal no preferential distribution of metals in reserves pits (Wojtanowicz et al., 1989). Because the migration rate of chromium out of unlined pits is only a few feet per decade (Campbell and Akers, 1990), reserves pits are not expected to be a major source of chromium contamination for the environment.

Regulations for the design and monitoring of reserves pits during and after drilling can vary significantly with location. Unlined pits are most commonly used for freshwater mud systems, while pits lined with an impermeable barrier are used for salt or oil-based mud systems. Following the completion of drilling of the well, the pits are eventually dewatered, covered with a few feet of soil, and abandoned.

For offshore applications, steel tanks are used as reserves pits. The solids, after being separated from the mud, are typically discharged into the sea, where they settle to the bottom around the drilling rig. In some areas, however, regulations require that any waste mud and cuttings be transported to shore for disposal.

2.1.5   Site Preparation

The preparation of drilling and production sites can cause local impact on the environment, including erosion, soil compaction, and sterilization. The development of a drilling site involves the construction of roads to the site and a level surface at the site. This construction can cause erosion. Erosion control measures like hay bales, silt fences, riprap, and mulching can be used. Environmentally sound construction methods are also required, such as slope controls, terracing, wing ditches, and diversion barriers.

The heavy equipment used to prepare a site can compact the soil, preventing water and nutrients from flowing through the pore system. This retards root development in plants and limits site restoration after abandonment. Depending on the site, it may take decades to recover (Powter, 1990). The level of compaction and its effects on plant growth depend on soil type and particle size distribution. To date, no good correlation has been developed to predict the effect of soil compaction on plant growth. Freeze/thaw and wetting/drying cycles have shown to be ineffective in loosening compacted soil and restoring normal water/air circulation.

Drilling sites are often sterilized with herbicides to prevent plant growth around the well and along rights-of-way. This reduces fire hazards and improves aesthetic appearance, particularly where weeds are prevalent. Depending on the herbicide and concentration used, however, treated areas can remain devoid of vegetation for many years. Often an excessive amount has been used to ensure long-term vegetation control with one application. When this occurs, the site becomes a potential source of contamination through surface runoff and wind dispersion to adjoining land. Bromacil and tebuthiuron have commonly been used as sterilization chemicals. These herbicides can become inactive by applying charcoal to the site at abandonment (Powter, 1990).

2.2 PRODUCTION

The production of oil and gas generates a variety of wastes. The largest waste stream is produced water, with its associated constituents. This section reviews both the production process and the wastes that are generated during production.

2.2.1 Overview of Production Processes

For the oil (or gas) to be produced, a pressure gradient must be established in the formation on the pore level. This pressure gradient then forces oil from one pore to the next, and ultimately to the production well. There are two basic ways for such a pressure gradient to be established. First is to have a production well with a lower pressure than that of the surrounding formation. This will cause oil to flow to the well, where it can be produced. Second is to increase the pressure in some parts of the formation by injecting fluids. This will force oil to flow away from the injection wells to lower pressure production wells. In many reservoirs, a combination of low pressure at the production well coupled with a high pressure at an injection well are used.

During production, both water and formation solids are commonly produced with oil and gas. The produced materials are passed through separation equipment, where the density differences between the produced materials are used to separate them.

The first stage of separation normally occurs in a free water knock-out. This consists of a large tank that allows time for the bulk oil, gas, and water phases to separate. These tanks are also called wash tanks, settling tanks, and gun barrels. The output streams from this equipment consists primarily of gas, water with some oil, and oil with some water. Solids either settle to the bottom of the tank or are carried along with the water stream. The performance of these separators has been reviewed by Powers (1990 and 1993) and the American Petroleum Institute (1990a).

The liquid streams exiting the free water knockout are generally in the form of an emulsion. These emulsions normally require additional treatment. Emulsions can be broken by adding demulsifiers (chemicals that cause the water drops to coalesce), by heating the emulsion, by passing an electrical current through the emulsion (Fang et al., 1991), or with combinations of these processes. These processes break the emulsion, allowing the droplets to grow and settle in the gravitational field. This settling is driven primarily by buoyancy and impeded by viscous drag, as described by Stokes law. Chemicals used to break emulsions include surfactants, alcohols, and fatty acids.

The efficiency of the separations equipment in breaking emulsions depends on the droplet size and density difference between the oil and water. Small droplets are much more difficult to separate. The droplet size depends on the interfacial tension between the oil and water and the shear history of the fluid. If the fluid flows through many shearing devices at high velocity, e.g., chokes, valves, or pumps, the oil can be shorn into smaller and smaller droplets. Emulsions are stabilized by many of the treatment chemicals added to the production stream, making separations even more difficult.

The hydrocarbon levels in the produced water after exiting demulsification equipment may still be too high for unrestricted discharge. Advanced water treatment methods are available that can further lower the hydrocarbon levels. These advanced methods are discussed in Chapter 6.

2.2.2 Produced Water

The largest volume waste stream in the upstream petroleum industry is produced water. For mature oil fields, the volume of produced water can be several orders of magnitude greater than the volume of produced oil. The environmental impact of produced waters arise from its chemical composition. Produced water contains dissolved solids and hydrocarbons (dissolved and suspended), and is depleted in oxygen.

Dissolved Solids

Most produced water contains a variety of dissolved solids. The most common dissolved solid is salt (sodium chloride). Salt concentrations in produced water range between a few parts per thousand to hundreds of parts per thousand (ppt). For comparison, seawater contains 35 parts per thousand.

In addition to salt, many produced waters also contain high levels of calcium, magnesium, and potassium, with lower amounts of aluminum, antimony, arsenic, barium, boron, chromium, cobalt, copper, gold, iron, lead, magnesium, manganese, nickel, phosphorus, platinum, radon, radium, silicon, silver, sodium, strontium, tin, uranium, and vanadium. The concentrations of seven major heavy metals found in produced water in the Gulf of Mexico are summarized in Table 2-6. Lead, nickel, chromium, zinc, nickel, and copper were found to have the highest concentrations (Stephenson, 1992). Produced water also contains low levels of naturally occurring radioactive materials. Radioactive materials are discussed below.

Hydrocarbons

Produced water normally contains dissolved and suspended droplets of hydrocarbons and other organic molecules that are not removed by the separations equipment. Hydrocarbon effluent concentrations vary widely with equipment used. The majority of the hydrocarbon concentrations in produced water from the Gulf of Mexico are between 10 and 30 mg/L, with virtually all levels less than about 100 ppm (Burke et al., 1991; Stephenson, 1992). The current U.S. Environmental Protection Agency limits for the discharge of hydrocarbons in water

Table 2-6
Heavy Metals Concentrations in Produced Water

Metal	Average Concentration (micrograms/L)	Standard Deviation (micrograms/L)
Cadmium	27	12
Chromium	186	68
Copper	104	180
Lead	315	670
Nickel	192	307
Silver	63	17
Zinc	170	253

Source: from Stephenson, 1992.
Copyright SPE, with permission.

for the *best available technology* (BAT) are 29 mg/L on a monthly average and 42 mg/L for a daily maximum. Like all regulatory targets, these numbers are subject to change.

The concentrations of dissolved hydrocarbons in produced water depends on the solubility of the hydrocarbon. For discharges in the Gulf of Mexico, dissolved hydrocarbon concentrations for phenols, benzene, and toluene were found to be between 1,000 and 6,000 micrograms/L, while the concentrations of high molecular weight hydrocarbons was considerably lower (Stephenson, 1992). These data are summarized in Table 2-7.

Oxygen Depletion

Produced water is invariably oxygen depleted. If discharged, oxygen depleted water can impact fauna requiring dissolved oxygen for respiration. Oxygen depletion can be a problem for discharge in shallow estuaries and canals, particularly if the produced water forms a layer along the bottom because of its higher density. This dense layer would be isolated from the atmosphere, limiting its contact with

Table 2-7
Dissolved Hydrocarbon Concentrations in Produced Water

Hydrocarbon	Average Concentration (micrograms/L)	Standard Deviation (micrograms/L)
Gas Production		
Phenols	4,743	5,986
Benzene	5,771	4,694
Toluene	5,190	4,850
C_2 Benzene	700	1,133
Oil Production		
Phenols	1,049	889
Benzene	1,318	1,468
Toluene	1,065	896
C_2 Benzene	221	754
Naphthalene	132	161
Other PAHs	7	18

Source: from Stephenson, 1992.
Copyright SPE, with permission.

oxygen. Oxygen depletion is normally not a problem for discharge in deep water or in high-energy environments because of rapid dilution of the produced water in the surrounding environment.

2.2.3 Production Chemicals

Produced water is responsible for a variety of problems in oilfield operations. The most common problems are emulsions, corrosion, scale, microbial growth, suspended particles, foams, and dirty equipment. A variety of chemicals are often added to the water to avoid those problems.

Emulsion Breakers

As previously discussed, produced water often consists of an oil-in-water emulsion. Chemicals are commonly used to lower the electrostatic forces on the oil droplets to allow them to coalesce into larger droplets. Common chemicals used for this purpose include surfactants, alcohols, and fatty acids.

Corrosion Inhibitors

Produced water can be very corrosive to production equipment. Corrosion is caused primarily by the presence of dissolved oxygen, carbon dioxide, and/or hydrogen sulfide gases. A detailed discussion of corrosion is given by Jones (1988).

Although produced water is initially oxygen depleted, oxygen can enter the produced fluid stream as a result of agitation during pumping or by atmospheric diffusion in holding tanks and surface impoundments. The oxygen content of water can be minimized by designing the system to exclude oxygen contact with the water. Carbon dioxide and hydrogen sulfide can occur naturally in the formation and be produced with the water. Carbon dioxide forms carbonic acid, which lowers the pH and increases the corrosivity of the water. Hydrogen sulfide corrosion can occur as a result of bacterial action on sulfates and is more often a surface or near surface phenomenon.

Complex inorganic salts like sodium chromate (Na_2CrO_4), sodium phosphate (Na_3PO_4), and sodium nitrite ($NaNO_3$) are also effective in slowing oxygen corrosion, particularly in high pH environments.

Sodium chromate, however, adds chromium to the produced water. Sodium hexametaphosphate ($Na_6P_6O_{18}$) is used in cooling and boiling water treatment. Zinc salts of organic phosphonic acids and sodium molybdate (Na_2MoO_4) have also been used for corrosion control. Zinc-based inhibitors are less toxic than chromates and should be used if possible. Organic anionic inhibitors, such as sodium sulfonates and sodium phosphonates, are also used in cooling waters and antifreeze. Current regulations may limit the use of some corrosion inhibitors.

Hydrogen sulfide can be removed from produced fluids with a zinc scavenger. Zinc carbonate ($ZnCO_3$-$Zn[OH]_2$) is widely used. This chemical reacts with hydrogen sulfide, producing insoluble zinc sulfide (ZnS).

For water injection systems, oxygen causes the largest problems with corrosion. Oxygen can be removed from water by stripping it with an inert gas, such as natural gas, steam, or flue gas, by vacuum deaeration, or by chemical treatment. Oxygen scavengers include sodium sulfite ($NaSO_3$), sodium bisulfite ($NaHSO_3$), ammonium bisulfite (NH_4HSO_3), sulfur dioxide (SO_2), sodium hydrosulfite ($Na_2S_2O_4$), and hydrazine (N_2H_2).

Cathodic protection can be used for external corrosion of casing and pipes and for internal corrosion of tanks. Both internal and external surfaces of surface equipment can sometimes be protected with liners to prevent corrosion. These liners can be hydrocarbon, plastic, metal, ceramic, or cement based.

Scale Inhibitors

The dissolved solids in produced water are normally in thermo-dynamic chemical equilibrium with the downhole conditions. As water is produced, however, its temperature and pressure are lowered, altering the chemical equilibrium. One common result of this altered chemical equilibrium is the precipitation of inorganic salts in production equipment, i.e., scale. Scale can plug production equipment, rendering it useless. Scale is commonly composed of calcium, strontium, and barium sulfates, as well as calcium carbonate. A more complete discussion of scaling is given by Jones (1988).

Scale can be inhibited by organic phosphate esters of amino-alcohols, phosphonates, or acrylic acid type polymers (sodium polyacrylate polymers). These chemicals adsorb onto the crystal nuclei when scale first precipitates and prevent further growth. Altering the design of the production system may also minimize the probability of a solution reaching a saturated state and forming scale in critical flow paths.

Because some oxygen scavengers can produce sulfates which can react with calcium, barium, and strontium to produce scale, the addition of oxygen scavengers where scaling may be a problem should be minimized.

A problem related to scale formation is the precipitation of hydrocarbon solids (paraffin) in production tubing and equipment. Paraffin precipitation occurs when the temperature and pressure of the crude oil no longer allow paraffin to remain dissolved in the oil. Various organic additives are used to inhibit paraffin deposition.

Biocides

Microbial growth (bacteria) in produced water can produce hydrogen sulfide gas by the chemical reduction of sulfates. Dissolved hydrogen sulfide gas makes produced gas highly corrosive. In addition to causing corrosion, the presence of the bacteria themselves can impact production operations. Bacterial fouling of equipment and degradation of hydrocarbons can occur. Pads or mats of bacteria, iron sulfide, and degraded oil can be formed at the oil/water interface in tanks and separators, rendering them less effective.

To minimize these problems, biocides are often added to the produced water to inhibit microbial growth. Surfactants can also be added to mobilize the microorganisims and make them more susceptible to the biocide. Bacteria are rarely completely killed using biocides, so long-term treatment is usually required once a system is contaminated. Biocides used include aldehydes, quaternary ammonium salts, and amine acetate salts. Chlorine compounds are used as biocides in municipal drinking water systems.

Coagulants

Produced water often contains various amounts of produced solids. While most of these solids are separated in surface settling tanks, very small solids (clay particles) may remain suspended in the water. Coagulants and flocculants can be added to the produced water stream to agglomerate these fine particles and allow them to settle.

Coagulants commonly include polyamines and polyamine quaternary ammonium salts.

Foam Breakers

Some crude oils generate a foam during production. This foam inhibits the separation of the oil, water, and solids in the production equipment. Although not commonly needed, foam breakers are available. Foam breakers include silicones, polyglycol esters, and aluminum stearate.

Surfactants

Surfactants (detergents) are regularly used to wash equipment and decks on offshore rigs. These surfactants commonly include alkyl aryl sulfonates and ethoxylated alkylphenols.

2.2.4 Well Stimulation

The oil and gas production rate of many wells is restricted by a low permeability around the wellbore. To increase the production rate, the permeability is often increased by stimulation. The two most common forms of stimulation are acidizing and hydraulic fracturing.

Acidizing

Acids are used to dissolve acid-soluble materials around the wellbore to increase the formation's permeability. These acid-soluble materials can include formation rocks and clays, as well as any materials added during drilling. A variety of inorganic and organic acids can be used, depending on the formation. These acids include hydrochloric, formic, acetic, and hydrofluoric. Additives are also required to optimize the process.

The most widely used acid is hydrochloric acid. Its main application is in low permeability carbonate reservoirs. The major reaction products produced during acidizing are carbon dioxide, calcium chloride, and water. Spent acid returned from a well has a high chloride content. The principal disadvantage of hydrochloric acid is its corrosivity on tubulars, particularly at temperatures above about 250°F.

Hydrofluoric acid is used to stimulate wells in sandstone formations. It is normally used in a mixture of hydrochloric or formic acids, and is used primarily to dissolve clays and muds. The reaction products are various forms of fluorosilicates. Like hydrochloric acid, it is highly corrosive.

Formic acid is a weak organic acid that is used in mixtures during stimulation. Formic acid is commonly used as a preservative. It is relatively noncorrosive and can be used at temperatures as high as 400°F.

Acetic acid is used to dissolve carbonate materials, either separately or in combination with hydrochloric or formic acid. It is a slowly reacting acid that can penetrate deep into the formation and is useful for high-temperature applications. Reaction products are calcium, sodium, or aluminum acetates. Acetate salts have minimal environmental impact. Like other organic acids, acetic acid has a relatively low corrosivity.

To prevent acids from damaging or destroying tubulars from corrosion, corrosion inhibitors are normally used. Many commercially available inhibitors are complex mixtures of organic compounds, including thiophenols, nitrogen heterocyclics, substituted thioureas, rosin amine derivatives, acetylenic alcohols, and arsenic compounds. Most corrosion inhibitors are retained in the reservoir, so very little is returned with the spent acid.

Highly reactive acids can react immediately with the formation. Because the benefits of an acid are maximized if the acid is allowed to penetrate deep into the formation before being spent, additives to reduce the reaction rate are used. A common way to retard the reaction rate is to emulsify the acid before injection, with the continuous phase being the additive. Emulsions retard the reaction rate by physically limiting the access of the acid to the formation. Commonly used additives include salts, alcohols, aromatic hydrocarbons, and other surfactants. Gelling agents, like xanthan gum and hydroxyethyl cellulose, alcohols, acrylic polymers, aliphatic hydrocarbons, and amines, are also used. Retarders such as alkyl sulfonates, alkyl amines, or alkyl phosphonates are also used to reduce the reaction rate by forming hydrophobic films on carbonate surfaces.

During production, the spent acid returning to the surface may become emulsified with crude oil. These emulsions can be stabilized by the fines released during acidizing. To prevent such emulsions from forming, demulsifiers (surfactants) can be used. Common demulsifiers include organic amines, salts of quaternary amines, and polyoxy-ethylated alkylphenols. Glycol ether can be used as a mutual solvent for both spent acid and oil.

Wettability agents are used to alter the relative permeability of emulsions during acidizing and to change the wettability back when acidizing is complete. The objective of such wettability changes is to lower the injection pressure by maximizing the relative permeability of the emulsion during injection and to maximize the subsequent production rate by maximizing the relative permeability of oil after acidizing. Wettability is changed by the use of surfactants such as ethylene glycol monobutyl ether, methanol, 2-butoxy ethanol, or fluorocarbons.

To lower the pumping pressure during acidizing, friction reducers are used with acid to reduce its viscosity. Friction reducers allow a higher injection rate for a given pump size or allow a smaller pump for a given injection rate. Friction reducers are normally organic polymers that convert Newtonian acid to shear-thinning, non-Newtonian fluid.

Solvents can be used as a preflush with acid to clean oil sludges and paraffin off of formation particles so they can be better contacted by the acid. These solvents normally have a high alcohol content, e.g., methanol or isopropanol.

Because the local permeability in a formation can vary significantly, the acid injection profile may not be uniform. To modify the injection profile and provide a more uniform acidization, fluid loss and diverting additives like benzoic acid flakes, naphthalene flakes (mothballs), rock salt, silica flour, or polymers can be used.

Even when corrosion inhibitors are used, some iron compounds will be dissolved into the acid and carried into the formation. In some cases, this iron can precipitate in the formation, reducing its permeability. Complexing agents, like citric, lactic, acetic, and gluconic acids, or derivatives like ethylene diamine tetracetic acid (EDTA) and nitrilo triacetic acid (NTA) can be used to inhibit the precipitation of iron.

Hydraulic Fracturing

Hydraulic fracturing increases the permeability around a wellbore by creating a high permeability channel from the wellbore into the formation. During hydraulic fracturing, fluids are injected at a rate high enough so that the fluid pressure in the wellbore exceeds the tensile strength of the formation, rupturing the rock.

The most commonly used base fluid for hydraulic fracturing is water. Water is inexpensive and inflammable. Various hydrocarbons can also be used as a base fluid, particularly where surface freezing may occur. Acid is also occasionally used when a combination of acidizing and hydraulic fracturing is desired. Liquefied gases, such as carbon dioxide or liquefied petroleum gases, can also be used, particularly to fracture gas wells. The use of a liquid base fluid in gas wells can reduce the gas production rate by lowering the gas relative permeability.

After fracturing, the fluid pressure in the fracture drops when the well is placed back on production. This allows the fracture to close. To keep the fracture open during production, solids are injected with the base fluid to fill the fracture and prop it open. Materials used for proppants include sand, aluminum pellets, glass beads, walnut shells, and plastic beads.

To lower the pump size required to fracture the rock, additives are used to increase the viscosity of the fracturing fluid to enhance its proppant-carrying capability. To viscosify the water-based fracture fluids, polymers such as guar or xanthan gum, cellulose, or acrylics can be used. These polymers are frequently cross-linked with metal ions like boron, aluminum, titanium, antimony, or zirconium to further enhance their viscosity. To viscosify the oil-based fracture fluids, aluminum phosphate esters are commonly used. Surfactants are also occasionally used to create a liquid-air foam or oil-water emulsion to be used as the fracture fluid. To prevent degradation of many gels at high temperatures, stabilizers like methanol and sodium thiosulfate can be added.

Most polymers and cross-linkers operate in a solution having an optimum pH. For fluids needing a low pH, buffers of acetic, adipic, formic, or fumeric acids can be used. For fluids needing a high pH, sodium bicarbonate or sodium carbonate can be used.

Many formations have sensitive clays that swell during water injection from the exchange of small cations inside the clays with larger water molecules. Swelling clays plug the pores, limiting fluid flow. Clay minerals can also break loose and migrate through the pore network to lodge in pore throats and limit fluid flow. Clay stabilizers are often used to prevent such damage. Temporary stabilization methods include adding salts to the fluids to minimize exchange of water molecules with the cations in the clays. These salts are then returned when the well is placed on production. Salts used as temporary stabilizers include sodium chloride, potassium chloride, calcium chloride, and ammonium chloride. Permanent stabilizers, such as quaternary amines and inorganic polynuclear cations like zirconium oxychloride or hydroxyaluminum, bond to the clay surfaces to stabilize them. Permanent stabilizers remain in the formation and are not removed with produced fluids.

When the viscosity of the fracture fluid is increased, the pressure drop in the pipe is also increased from friction. This results in a higher pressure at the pump, but a lesser increase in pressure at the formation face where it is needed. To suppress the pressure drop in the pipe, high molecular weight polymers can be added to the fracture fluid. These polymers suppress turbulence, keeping the flow in the pipe laminar and lowering the friction losses.

A related method for reducing the pressure drop in the pipe is to use a cross-linking polymer that has a slow gelling time. The cross-linkers are added to the polymer at the wellhead just prior to injection. The mixing is timed so that the gel reaches its maximum strength when it reaches the formation face. This causes the maximum fluid pressure at the formation face and minimizes the pressure drop down the pipe.

The polymers used to alter the viscosity of fracturing fluids are subject to bacterial degradation. Bactericides, such as glutaraldehyde, chlorophenates, quaternary amines, and isothiazoline, are often added to control the level of bacteria.

To control fluid loss into high permeability zones, fluid loss additives can be added to fracture fluids. These solids include silica flower, granular salt, carbohydrates, and proteins for water-based fluids and organic particulates such as wax, pellets, or naphthalene granules for oil-based fluids. Another popular fluid loss method is to use an oil-in-water emulsion. This causes two-phase flow through the filter cake along the fracture wall, lowering the relative permeability of the water through the filter cake.

After a fracture has been created, breakers are used to lower the gel viscosity so the fracture fluids can be easily removed from the fracture and not inhibit subsequent production. A common breaker for water-based fracture fluids are peroxydisulfates. Altering the pH by adding acids or bases is a common way to break oil-based fracture fluids.

Following hydraulic fracturing, sand is often produced from the wells. To minimize sand production, chemicals that physically stabilize the sand around the wellbore can be injected. These chemicals include plastics like phenol formaldehyde and epoxy resins, together with alcohol solvents and special refined oils.

2.2.5 Natural Gas Production

As natural gas flows from the ground, it contains a variety of impurities that must be removed before it can be sold. These impurities are primarily water vapor, carbon dioxide, and hydrogen sulfide. The process of removing hydrogen sulfide and carbon dioxide is called *sweetening.*

Natural gas also contains fluids like propane, butane, and ethane, which can be separated from the gas by liquefaction. These natural gas liquids are more valuable and can be sold at higher prices. Other materials contained in the gas stream include produced water, pigging materials for the pipelines, filter media, fluids from corrosion treatment, and solids like rust, pipe scale, and produced sand. Cooling water and used lube oils and filters from compressors are also generated during gas treatment (American Petroleum Institute, 1989).

Natural gas is separated from produced solids and liquids by gravitational forces in separators. Natural gas liquids are separated from the lower molecular weight components by compression, absorption, and refrigeration.

Water vapor is removed from natural gas by contact with liquid or solid desiccants. Liquid desiccants include triethylene glycol, ethylene, and diethylene. Solid desiccants include towers filled with alumina, silica gel, silica-alumina beads, or molecular sieves. The water is subsequently removed from the desiccant by heat regeneration, and the desiccant is reused. The desiccation processes can generate wastes of glycol-based fluids, glycol filters, condensed water, and solid dessicants. These materials may contain low levels of hydrocarbons and treating chemicals. Benzene and other volatile aromatics can dissolve in glycols and be subsequently emitted when the glycol is being regenerated for reuse.

Carbon dioxide and hydrogen sulfide are removed from natural gas by contact with amines. The most common amines are diethanolamine (DEA) and monoethanolamine (MEA). Hydrogen sulfide can also be removed by contact with sulfinol, iron sponges (finely divided iron oxide in wood shaving carriers), and caustic solutions. Amines and sulfinol can be restored for reuse by heat regeneration, but iron sponges and caustic solutions are spent as the iron is converted to iron sulfide and other sulfur compounds. Other wastes generated when removing sweetening natural gas include spent amine, used filter media, and flared acid gas wastes. Sodium hydroxide is often added to the amine to prevent corrosion of equipment.

During sweetening, amine compounds are attacked by carbon dioxide and can break down. The solutions are filtered to remove the degradation products from the usable amine. The degradation products form toxic amine sludges that require treatment and disposal (Boyle, 1990).

During the production of natural gas, hydrates can form from the gas and water vapor. Hydrates are a slushy, ice-like substance that can plug the production tubing and equipment. Various chemicals, primarily methanol and ethylene glycol, are sometimes added to gas-producing wells to lower the freeze point of hydrates to inhibit their formation.

2.2.6   Other Operations

A variety of other operations associated with the production of oil and gas generate wastes that have the potential to impact the environment. These wastes include wastewater from cooling towers, water softening wastes, contaminated sediments, scrubber wastes, used filter media, various lubrication oils, and site construction wastes.

Cooling towers are used for a variety of processes during oil and gas production. The cooling water used in these towers often contains chrome-based corrosion inhibitors and pentachlorophenol biocides.

In many areas, produced water is reinjected into the reservoir to assist hydrocarbon recovery. Unfortunately, the level of dissolved solids, particularly hardness ions (calcium and magnesium), is often too high to be used because they readily precipitate and can plug the formation. Thus, before produced water can be reinjected, it must be softened to exchange the hardness ions with softer ions, e.g., sodium.

The most common way to soften produced water is through ion exchange. There are two major ion exchange resins (substrates) that are commonly used: strong acid resins, using sulfonic acid, and weak acid resins, using carboxylic acid. Strong acid resins can be regenerated simply by flushing with a concentrated solution of sodium chloride. Weak acid resins, however, must be regenerated by flushing with a strong acid-like hydrochloric or sulfuric and then neutralizing it with sodium hydroxide.

During oil production, sand and shale sediments are often produced with the oil. These sediments are separated out in the surface equipment. They normally collect in tank bottoms and must be periodically removed. These solids are normally mixed with oil, forming a sludge. Sediments can also be contaminated with oil and other materials from spills and leaks from equipment.

The hydrocarbon content of oil-contaminated sediments can exceed 4% by weight (Deuel, 1990). These sediments may also contain heavy metals or hydrogen sulfide (Brommelsiek and Wiggin, 1990). Total heavy metal concentrations in produced solids are generally low, as indicated in Table 2-8 (Cornwell, 1993). It is not known whether the differences in heavy metal concentrations for native soils in Alaska, shown in Table 2-2, and for produced solids, shown in Table 2-8, are from production activities or just natural variations in geology.

To remove the suspended solids that are not removed by settling, produced fluids are often passed through filters. The filter media must be frequently replaced or backwashed. The filled filters or filter backwash must be disposed.

The operations of much of the oilfield equipment, including stuffing boxes, compressors, and pumps, requires lubrication oil. As this oil is used, it changes its composition, making it potentially unsuitable for future use. The used lube oil must be replaced with fresh oil, and the used oil must be disposed of.

In areas where lease crude is burned, e.g., where steam is injected to recover oil, the combustion gases may need to be scrubbed to remove pollutants like sulfur dioxide. One way to remove sulfur dioxide from combustion gases is to bubble it through aqueous solutions containing caustic chemicals like sodium hydroxide or sodium carbonate. Sulfur dioxide dissolves into water, forming sulfuric acid, which is neutralized by the caustic. Another form of scrubber uses various amines. Typical wastewaters can have very high levels of dissolved solids, as indicated in Table 2-9 (Sarathi, 1991).

In cold climates, like Alaska's North Slope, methanol is used for freeze protection of equipment. It is used to protect pipelines, shut-in

Table 2-8
Heavy Metal Concentrations
of Produced Solids

Metal	Total (ppm, wt)
Antimony	11
Arsenic	105
Barium	326
Beryllium	<1
Cadmium	<1
Chromium	93
Cobalt	10
Copper	64
Lead	22
Mercury	5
Molybdenum	16
Nickel	176
Selenium	<1
Silver	1
Thallium	17
Vanadium	27
Zinc	214
Fluoride	76

Source: from Cornwell, 1993.
Copyright SPE, with permission.

water injection wells, and as a component of water-based hydraulic fracturing fluids.

2.2.7   Radioactive Materials

Many drilling sites and production facilities have radioactive materials associated with them. Some of these radioactive materials, primarily radioactive tracers or logging tools, are deliberately brought to the site for use, while other materials are naturally occurring and are called *naturally occurring radioactive materials* (NORM).

Radioactive Sources and Tracers

Radioactive sources are used primarily during logging with wireline tools. Both gamma ray and neutron sources are available. These

Table 2-9
Composition of Scrubber Wastewater

Constituent	Concentration (ppm)
Aluminum	0.4
Bicarbonate	31,183
Boron	20.8
Calcium	11.2
Carbonate	0
Chloride	2,237
Copper	0.5
Fluoride	5.2
Iron	32
Magnesium	0.43
Manganese	0.63
Nitrate	0.5
Phosphate	0.6
Potassium	101
Sodium	53,000
Sulfate	79,013
Sulfur Dioxide	420
Zinc	5.3
Total Dissolved Solids	148,438

Source: from Sarathi, 1991.
Copyright SPE, with permission.

sources are sealed within the logging tools and are normally not a problem.

Radioactive tracers are commonly used in injection wells to determine points of fluid entry into the formation (injection profile), hydraulic fracture height, and/or fluid leaks in the cement behind casing. The tracer is injected into the wellbore and a gamma ray detector is then logged through the well to determine depths at which the radioactivity is high. Commonly used radioactive tracers for liquid phase measurements include antimony-124 (as antimony oxide), iridium-192 (as potassium hexachloroiridate), scandium-46 (as scandium chloride), and iodine-131 (as sodium iodide). Krypton-85 has been used as a vapor phase tracer. Radioactive proppant is also used during hydraulic fracturing to monitor the location of the fracture behind casing. Radioactive proppants typically use the same isotopes that are used for liquid phase tracers.

Naturally Occurring Radioactive Materials

Naturally occurring radioactive materials (NORM) are found virtually everywhere on the earth, including ground and surface waters (Judson and Osmond, 1955). During the production of oil and gas, radioactive materials that naturally occur within the earth can be coproduced. Although the concentrations of NORM are usually very low, these materials can be concentrated during production; the concentrated levels can become high enough to cause a health hazard if improperly managed.

There are four radionuclides most commonly found in NORM in the upstream petroleum industry: radium-226, radium-228, radon-222, and lead-210. Radium-226 is probably the nuclide with the greatest potential for environmental impact for the petroleum industry. Other radioactive materials are also found, but in significantly lower amounts.

Radium (both 226 and 228) is highly soluble and is produced as a dissolved solid with the produced water. The levels of radium in produced water vary significantly. Although most wells do not produce significant amounts of NORM, typical concentrations in wells having NORM have been reported to vary between 1–2,800 picocuries per liter (pCi/l). Much higher concentrations, however, have also been reported (St. Pe et al., 1990; Miller et al., 1990; Snavely, 1989; Stephenson, 1992). In comparison, the natural radium levels in surface waters are typically less than 1 pCi/l. Drinking water standards for radioactive materials are typically 5 pCi/l, and discharge standards for open water are 30 pCi/l, although these regulatory limits can vary.

Radium is coprecipitated with barium, calcium, and strontium sulfate as scale in tubulars and surface equipment during production. This concentrates the radium and makes the scale radioactive. Radium can also be concentrated in various production sludges through its association with solids in the sludge. NORM concentrations of several hundred thousand pCi/gm have been found in scale in piping and surface equipment. Concentrations in excess of 8,000 pCi/gm have been measured in the soil at pipe cleaning yards (Carroll et al., 1990). The presence of NORM, however, can be easily identified with gamma ray detectors.

Radon-222 is a naturally occurring gas that is found in some produced water and natural gas liquids. This gas comes out of solution as the pressure is reduced during production. Because it is a gas, it normally is not concentrated in sufficient quantities to cause environmental impact, although it can be temporarily concentrated in low-lying areas.

Lead-210 is of particular concern to the natural gas liquids industry (Gray, 1993). When lead-210 is formed, it precipitates on equipment surfaces, forming an extremely thin layer of radioactive film.

Although significant levels of NORM have been seen at some production operations, it is not normally encountered at drill sites. The drilling process does not provide a way for significant concentrations of NORM to accumulate.

2.3   AIR EMISSIONS

A wide variety of air pollutants are generated and emitted during the processes of finding and producing petroleum. These air pollutants include primarily oxides of nitrogen (NO_x), volatile organic compounds (VOCs), oxides of sulfur (SO_x), and partially burned hydrocarbons (like carbon monoxide and particulates). Dust from construction and unpaved access roads can also be generated.

Volatile hydrocarbons, including aromatics, are emitted during the regeneration of glycol from natural gas dehydration (Grizzle, 1993; Thompson et al., 1993). Halon gases are used at many drilling and production sites for fire suppression. These gases have been identified as an ozone-depleting chlorofluorocarbon (CFC), and their use releases them to the atmosphere.

2.3.1   Combustion

The largest source of air pollution in the petroleum industry is the operation of the internal combustion engines used to power drilling and production activities, such as drilling rigs, compressors, and pumps. These engines can be powered by either natural gas or diesel fuel. The two primary pollutants emitted from these engines are oxides of nitrogen, primarily NO and NO_2, and partially burned hydrocarbons. The nitrogen oxides are commonly referred to as NO_x. During combustion, about 3.5 pounds of NO_x can be generated for each barrel of fuel burned.

Emissions of NO_x from petroleum industry operations in 1975 totaled 1.3 million U.S. tons. This level was about 11% of the total NO_x emissions from all stationary sources in the United States and

6% of the total emissions from all sources. About 46% of the NO_x emitted by the petroleum industry was from gas processing activities, 21% from production activities, and 22% from refineries. Crude oil transport emitted 5.2% of the petroleum industry NO_x, onshore drilling emitted 4.2%, and product transport emitted 0.9% (American Petroleum Institute, 1979).

NO_x is formed at high combustion temperatures when molecular oxygen dissociates into individual oxygen atoms. Atomic oxygen readily reacts with atmospheric nitrogen to form NO_x. Methods to limit the formation of NO_x include combustion modifications to lower the flame temperature during combustion and flue gas treatment to remove any NO_x that has formed. However, little can be done during drilling and production operations to lower NO_x emission, other than to purchase low NO_x generating equipment and operate it as recommended by the manufacturer.

Partially burned hydrocarbons are emitted during combustion when the fuel/air mixture is incorrect. The most common partially-burned hydrocarbons from internal combustion engines powered by natural gas are formaldehyde and benzene (Meeks, 1992). About 25 pounds of formaldehyde and 1.5 pounds of benzene can be generated per million cubic feet (MMcf) of fuel burned. For fuels containing benzene, ethylbenzene, toluene, or xylene (BETX), about 3% of those compounds will pass through the engine and be emitted.

Another major source of air pollutants is the operation of heater treaters, boilers, and steam generators. These types of equipment also emit NO_x and partially burned hydrocarbons like carbon monoxide. If a sulfur-bearing fuel is used, sulfur oxides, primarily SO_2 and SO_3 (referred to as SO_x), can also be emitted. For a crude oil having a sulfur content of 1.1%, about 7.5 pounds of sulfur will be released for every barrel of fuel burned. Table 2-10 shows the typical emission levels of an oil-fired steam generator operating at different levels (Sarathi, 1991). For reference, a steam generator operating at 50 million Btu/hr can inject steam into three to five wells. The data in this table were adjusted for 365 days of continuous operation.

2.3.2 Emissions from Operations

A number of operations at production facilities emit volatile materials into the air. Operations that can cause emissions include the use

Table 2-10
Typical Steam Generator Emission Levels

Operating Level (Million Btu/hr)	SO_2 (tons/year)	NO_2 (tons/year)	Particulates (tons/year)	Hydrocarbons (tons/year)
5	21	10.3	2.9	0.43
10	66	23	6.4	0.91
20	151	53	15	2.0
50	275	96	27	3.8

Source: adapted from Sarathi, 1991.
Copyright SPE, with permission.

of fixed roof tanks, wastewater tanks, loading racks, and casing gas from thermal recovery operations. A more detailed discussion of the emissions from a typical onshore oil and gas production facility is provided by Sheehan (1991) and Smith (1987).

During the operation of fixed roof tanks, volatile hydrocarbons can be emitted into the atmosphere. There are three major sources of emissions from these tanks: breathing losses, working losses, and flashing losses. Breathing losses arise from a change in vapor volume from changes in temperature and barometric pressure. Working losses are caused by changes in the tank's fluid level. Flashing losses occur when dissolved gas flashes to vapor from pressure drop changes between the tank and the production line. A detailed description on calculating emissions from fixed roof tanks has been prepared by the American Petroleum Institute (1991).

Open tanks, sumps, and pits can be sources of emissions for volatile hydrocarbons. The emission rates depend on the ambient temperature, surface area of the fluid exposed to the atmosphere, and composition of the hydrocarbon.

Another operational source of air emissions is the transfer of oil from tanks to trucks. These emissions occur when the vapors in the truck are displaced by the entering fluid.

During production from thermal recovery projects, hot fluids are produced at the production well. Hydrocarbon vapors, carbon dioxide, and various sulphur compounds can be produced with the oil or from the casing annulus. To prevent these gases from escaping into the atmosphere, they can be collected and processed in a *casing vapor recovery system* (Peavy and Braun, 1991). Such systems can remove

99% of the hydrocarbon vapors and 95% of the sulfur in the casing gas. Because of the sales value of the condensed hydrocarbon vapors, these systems can pay out within a few years.

2.3.3   Fugitive Emissions

Another source of air pollutants are the fugitive emissions of volatile hydrocarbons. These are hydrocarbons that escape from production systems through leaking components like valves, flanges, pumps, compressors, connections, hatches, sight glasses, dump level arms, packing seals, fittings, and instrumentation. Valves are usually the most common components that leak. These emissions generally result from the improper fit, wear and tear, and corrosion of equipment. Although the leak rate from individual components is normally small, the cumulative emissions from an oil field containing a large number of components can be significant.

A comprehensive study of fugitive hydrocarbon emissions from petroleum production operations revealed that an average of about 5% of all components in field locations leak (American Petroleum Institute, 1980). A breakdown of how often each type of component leaked is given in Table 2-11. Components in gas service have a leak rate that is about an order of magnitude higher than components in liquid service. The leak rate at offshore production facilities is significantly lower than at onshore facilities.

Table 2-11
Fugitive Emissions from Petroleum Operations Equipment

Component	Total Number Tested	% Leaking
Valve	25,089	8.4
Connection	138,510	3.4
Sightglass	676	1.3
Hatch	358	6.1
Seal packing	1,246	25.9
Diaphragm	1,643	19.4
Meter	92	5.4
Sealing mechanism	5,591	10.9
Total/Average	**173,205**	**4.7**

Source: American Petroleum Institute, 1980.
Reprinted by permission of the American Petroleum Institute.

Because of the cost of obtaining fugitive emission data, emission rates are typically measured carefully at only a few facilities. The data obtained are then normalized to the number and type of fittings to be used at other facilities. One such set of generic fugitive emission factors for a production facility that is based on the number of production wells and the gas/oil ratio is given in Table 2-12.

More accurate sets of fugitive emission factors can be based on the number of valves, connections, fittings, flanges, and similar equipment at a facility. The estimate for the total fugitive emissions would then be the sum of the average emissions from each piece of equipment (Schaich, 1991). Table 2-13 provides a list of average emission factors for various types of equipment.

Past studies indicate that emission factors such as those given in Table 2-13 can overestimate emissions by several orders of magnitude. A more accurate method of estimating fugitive emissions is to measure how many pieces of equipment are leaking and apply one set of fugitive emission factors to the components that are leaking and a second set to the components that are not leaking. A set of these generic fugitive emission factors is given in Table 2-14. In this table, a fitting is assumed to leak if the concentration measured by a hand-held analyzer is greater than 10,000 ppm-v (parts per million by volume).

If a more refined measurement of emission concentration at a piece of equipment is made, an even more accurate set of fugitive emission factors can be generated. One such set of factors for three emission ranges is given in Table 2-15. An even more refined approach would

(text continued on page 64)

Table 2-12
Generic Fugitive Emission Rates for Production Facilities

Number of Wells	Gas/Oil Ratio	Emission rate (lbm/well/day)
<10	<500	2.56
10–50	<500	1.44
>50	<500	0.09
<10	=>500	6.85
10–50	=>500	2.89
>50	=>500	4.34

Source: from Sheehan, 1991.
Copyright SPE, with permission.

Table 2-13
Generic Fugitive Emission Factors for Production Equipment

Equipment	Fluid	Emission Factor (kg/hr/source)
Valves	Gas	0.0056
	Light liquid	0.0071
	Heavy liquid	0.00023
Pump seals	Light liquid	0.0494
	Heavy liquid	0.0214
Compressor seals	Gas/vapor	0.228
Pressure relief devices	Gas/vapor	0.104
Flanges	All	0.00083
Open-ended lines	All	0.0017
Sampling connections	All	0.0150

Source: from Schaich, 1991.

Table 2-14
Fugitive Emission Factors Based on Leak Determination

Equipment	Service	Emission Factor: Leaking (kg/hr/source)	Emission Factor: Nonleaking (kg/hr/source)
Valves	Gas	0.0451	0.00048
	Light liquid	0.0852	0.00171
	Heavy liquid	0.00023	0.00023
Pump seals	Light liquid	0.437	0.012
	Heavy liquid	0.3885	0.0135
Compressor seals	Gas/vapor	1.608	0.0894
Pressure relief devices	Gas/vapor	1.691	0.0447
Flanges	All	0.0375	0.00006
Open-ended lines	All	0.01195	0.00150

Source: from Schaich, 1991.

Table 2-15
Fugitive Emission Factors for Three Leakage Rates

Equipment	Service	Emission Factor: 0–1,000 ppm (kg/hr/source)	Emission Factor: 1,001–10,000 (kg/hr/source)	Emission Factor: >10,000 (kg/hr/source)
Valves	Gas	0.00014	0.00165	0.0451
	Light liquid	0.00028	0.00963	0.0852
	Heavy liquid	0.00023	0.00023	0.00023
Pump seals	Light liquid	0.00198	0.0335	0.437
	Heavy liquid	0.0038	0.0926	0.3885
Compressor seals	Gas/vapor	0.01132	0.264	1.608
Pressure relief devices	Gas/vapor	0.0114	0.279	1.691
Flanges	All	0.00002	0.00875	0.0375
Open-ended lines	All	0.00013	0.00876	0.01195

Source: from Schaich, 1991.
Reproduced with permission of the American Institute of Chemical Engineers.

(text continued from page 61)

be to develop correlations between the emission rates and the concentrations measured by a hand-held detector. One set of such correlations is given in Table 2-16.

The biggest problem with using emission factors with measurements made with hand-held detectors is that the local concentration of emitted hydrocarbons varies considerably with local conditions. Conditions that can affect these measurements are wind speed, pressure in fitting, composition of hydrocarbon in fitting, and location of detector when taking the measurement.

2.3.4 Emissions from Site Remediation

Another source of air pollution is from the cleanup of petroleum contaminated sites. Many cleanup practices for hydrocarbons spilled on soil result in volatile hydrocarbons being emitted into the air and transported from the spill site. The most common hydrocarbon spilled that causes air pollution is gasoline. Models have been developed to estimate the pollutant levels associated with three types of soil cleanup technologies: soil extraction, vacuum extraction, and air stripping (U.S. Environmental Protection Agency, 1989).

Soil extraction is commonly used when contaminated soil is dumped in a pile to be treated and/or disposed of at a later date. When liquid gasoline and air are present in the soil, the concentration of volatile organic carbons (VOCs) will build until it reaches local equilibrium.

Table 2-16
Fugitive Emission Rates Based on Correlation

Source	Service	Equation
Valves	Gas/vapor	$Q = 3.766 \times 10^{-5.35} \, C^{0.693}$
Valves	Light liquid	$Q = 8.218 \times 10^{-4.342} \, C^{0.47}$
Pump Seals	All	$Q = 2.932 \times 10^{-5.34} \, C^{0.898}$
Flanges	All	$Q = 2.10 \times 10^{-4.733} \, C^{0.818}$

Q is the emission rate in lbm/hr.
C is the measured maximum concentration at the fitting in ppm-v.

Source: from Schaich, 1991.

The VOC and benzene levels are typically higher for this remediation method than for other methods, but have shorter durations of emission. Typical VOC emissions for a soil pile having an area of 2,000 ft^2 are between 50 and 200 lbm/hr, depending on the temperature. Benzene emissions for the pile typically range from 0.5 to 2 lbm/hr.

One way to extract the volatile hydrocarbon components in soil is by vacuum extraction. Vacuum extraction consists of drilling a well through the contaminated soil and pulling a vacuum in the well. The lower pressure forces air into the pile, and volatilized compounds are vacuumed with the air into the well and removed from the pile. Because soil is treated in place, vacuum extraction can be less expensive and less disruptive than other methods. Maximum emission rates tend to be under 50 lbm/hr for VOCs and under 2 lbm/hr for benzene. The duration of emissions tends to be on the order of weeks to months.

Volatile hydrocarbons can also be removed from contaminated water that has been pumped from the ground by air stripping. In this process, the contaminated water is allowed to trickle over packing material in an air stripping tower. Clean air is simultaneously circulated through the packing material. The volatile hydrocarbons vaporize into the air and are released to the atmosphere. The removal efficiency depends on the contaminant, but is typically 99% to 99.5%. Emissions of volatile hydrocarbons tend to be between 0.5 to 4 lbm/hr, with benzene releases between 0.1 and 0.5 lbm/hr. Although air stripping has the lowest emission levels of the three methods discussed here, it typically has the longest duration.

REFERENCES

American Petroleum Institute, "NO$_x$ Emissions from Petroleum Industry Operations," API Publication 4311, Washington, D.C., Oct. 1979.

American Petroleum Institute, "Fugitive Hydrocarbon Emissions from Petroleum Production Operations: Volumes I and II," API Publication 4322, Washington, D.C., March 1980.

American Petroleum Institute, "API Environmental Guidance Document: Onshore Solid Waste Management in Exploration and Production Operations," Washington, D.C., Jan. 1989.

American Petroleum Institute, "Monographs on Refinery Environmental Control-Management of Water Discharges: Design and Operation of Oil-Water Separators," API Publication 421, Washington, D.C., Feb. 1990a.

American Petroleum Institute, "Monographs on Refinery Environmental Control-Management of Water Discharges: The Chemistry and Chemicals of Coagulation and Flocculation," API Publication 420, Washington, D.C., Aug. 1990b.

American Petroleum Institute, "Manual of Petroleum Measurement Standards: Chapter 19—Evaporative Loss Measurement, Section 1—Evaporative Loss from Fixed-Roof Tanks," API Publication 2518, Washington, D.C., Oct. 1991.

Bleier, R., Leuterman, A. J. J., and Stark, C., "Drilling Fluids: Making Peace with the Environment," *J. Pet. Tech.,* Jan. 1993, pp. 6–10.

Boyle, C. A., "Management of Amine Process Sludges," Proceedings of the U.S. Environmental Protection Agency's First International Symposium on Oil and Gas Exploration and Production Waste Management Practices, New Orleans, LA, Sept. 10–13, 1990, pp. 577–590.

Brommelsiek, W. A. and Wiggin, J. P., "E & P Waste Management in the Complex California Regulatory Environment—An Oil and Gas Industry Perspective," Proceedings of the U..S. Environmental Protection Agency's First International Symposium on Oil and Gas Exploration and Production Waste Management Practices, New Orleans, LA, Sept. 10–13, 1990, pp. 293–306.

Burke, N. E., Curtice, S., Little, C. T., and Seibert, A. F., "Removal of Hydrocarbons From Oil Field Brines by Flocculation with Carbon Dioxide," paper SPE 21046 presented at the Society of Petroleum Engineers International Symposium on Oilfield Chemistry, Anaheim, CA, Feb. 20–22, 1991.

Campbell, R. E. and Akers, R. T., "Characterization and Cleanup of Chromium-Contaminated Soil for Compliance with CERCLA at the Naval Petroleum Reserve No. 1 (Elk Hills): A Case Study," paper SPE 20714 presented at the Society of Petroleum Engineers 65th Annual Technical Conference and Exhibition, New Orleans, LA, Sept. 23–25, 1990.

Candler, J., Leuterman, A., Wong, S., and Stephens, M., "Sources of Mercury and Cadmium in Offshore Drilling Discharges," paper SPE 20462 presented at the Society of Petroleum Engineers 65th Annual Technical Conference and Exhibition, New Orleans, LA, Sept. 23–25, 1990.

Carroll, J. F., Scott, H. D., Gunn, R. A., and O'Brien, M. S., "Naturally Occurring Radioactive Material Logging," paper SPE 20616 presented at the Society of Petroleum Engineers 65th Annual Technical Conference and Exhibition, New Orleans, LA, Sept. 23–26, 1990.

Clark, R. K., "Impact of Environmental Regulations on Drilling-Fluid Technology," *J. Pet. Tech.,* Sept. 1994, pp. 804–809.

Clark, R. K. and Almquist, S. G., "Evaluation of Spotting Fluids in a Full-Scale Differential-Pressure Sticking Apparatus," *SPE Drilling Engineering,* June 1992, p. 121.

Cornwell, J. R., "Road Mixing Sand Produced from Steam Drive Operations," paper SPE 25930 presented at the Society of Petroleum Engineers/ Environmental Protection Agency's Exploration and Production Environmental Conference, San Antonio, TX, March 7–10, 1993.

Deeley, G. M., "Use of MINTEQ for Predicting Aqueous Phase Trace Metal Concentrations in Waste Drilling Fluids," Proceedings of the U.S. Environmental Protection Agency's First International Symposium on Oil and Gas Exploration and Production Waste Management Practices, New Orleans, LA, Sept. 10–13, 1990, pp. 1013–1023.

Deuel, L. E., "Evaluation of Limiting Constituents Suggested for Land Disposal of Exploration and Production Wastes," Proceedings of the U.S. Environmental Protection Agency's First International Symposium on Oil and Gas Exploration and Production Waste Management Practices, New Orleans, LA, Sept. 10–13, 1990, pp. 411–430.

Deuel, L. E. and Holliday, G. H., "Reserve Pit Drilling Wastes—Barium and Other Metal Distributions of Oil Gas Field Wastes," paper SPE 20712 presented at the Society of Petroleum Engineers 65th Annual Technical Conference and Exhibition, New Orleans, LA, Sept. 23–25, 1990.

Enright, D. P. and Smith, F. M., "An Environmentally Safe Water-Based Alternative to Oil Muds," paper SPE/IADC 21937 presented at the Society of Petroleum Engineers 1991 Drilling Conference, Amsterdam, The Netherlands, March 11–14, 1991.

Fang, C. S., Tong, N. A. M., and Lin, J. H., "Removal of Emulsified Crude Oil from Produced Water by Electrophoresis," paper SPE 21047 presented at the Society of Petroleum Engineers International Symposium on Oilfield Chemistry, Anaheim, CA, Feb. 20–22, 1991.

Farrell, C. W., "Oilfield Process Stream Treatment by Means of Alternating Current Electrocoagulation," *Advances in Filtration and Separation Technology, Vol. 3: Pollution Control Technology for Oil and Gas Drilling and Production Operations,* American Filtration Society. Houston: Gulf Publishing Co.; 1991, pp. 186–207.

Friedheim, J. E. and Shinnie, J. R., "New Oil-Base Mud Additive Reduces Oil Discharged on Cuttings," paper SPE/IADC 21941 presented at the Society of Petroleum Engineers 1991 Drilling Conference, Amsterdam, The Netherlands, March 11–14, 1991.

Gray, P. R., "NORM Contamination in the Petroleum Industry," *J. Pet. Tech.,* Jan. 1993, pp. 12–16.

Grizzle, P. L., "Hydrocarbon Emission Estimates and Controls for Natural Gas Glycol Dehydration Units," paper SPE 25950 presented at the Society of Petroleum Engineers/Environmental Protection Agency's Exploration and Production Environmental Conference, San Antonio, TX, March 7–10, 1993.

Hoberock, L. L., "A Study of Vibratory Screening of Drilling Fluids," *J. Pet. Tech.,* Nov. 1980, pp. 1889–1900.

Jones, L. W., *Corrosion and Water Technology.* Tulsa: OGCI Publications, 1988.

Judson, S. and Osmond, J. K., "Radioactivity in Ground and Surface Water," *American Journal of Science,* Vol. 253, Feb. 1955, pp. 104–116.

Lal, M., "Economics and Performance Analysis Model for Solids Control," paper SPE 18037 presented at the 1988 Society of Petroleum Engineers Annual Technical Conference and Exhibition, Houston, TX, Oct. 2–5, 1988.

Lal, M. and Hoberock, L. L., "Solids-Conveyance Dynamics and Shaker Performance," SPE Drilling Engineering, Dec. 1988, pp. 385–392.

Lal, M. and Thurber, N. E., "Drilling Wastes Management and Closed-Loop Systems," Drilling Wastes: Proceedings of the 1988 Intl. Conference on Drilling Wastes, Calgary, Canada, April 5–8. New York City: Elsevier Applied Science, 1989, pp. 213–228.

Leuterman, A. J. J., Jones, F, V., and Candler, J. E., "Drilling Fluids and Reserve Pit Toxicity," *J. Pet. Tech.,* Nov. 1988, pp. 1441–1444.

McDonald, H. B., "Thread Compounds + Environment = Change," *J. Pet. Tech.,* July 1993, pp. 614–616.

Meeks, H. N., "Air Toxics from Gas-Fired Engines," *J. Pet. Tech.,* July 1992, pp. 840–845.

Miller, H. T., Bruce, E. D., and Scott, L. M., "A Rapid Method for the Determination of the Radium Content of Petroleum Production Wastes," Proceedings of the U.S. Environmental Protection Agency's First International Symposium on Oil and Gas Exploration and Production Waste Management Practices, New Orleans, LA, Sept. 10–13, 1990, pp. 809–820.

National Research Council, *Oil in the Sea: Inputs, Fates, and Effects.* Washington, D.C.: National Academy Press, 1985.

Peavy, M. A. and Braun, J. E., "Control of Waste Gas From a Thermal EOR Operation," *J. Pet. Tech.,* June 1991, pp. 656–661.

Peresich, R. L., Burrell, B. R., and Prentice, G. M. "Development and Field Trial of a Biodegradable Invert Emulsion Fluid," paper SPE/IADC 21935 presented at the Society of Petroleum Engineers 1991 Drilling Conference, Amsterdam, The Netherlands, March 11–14, 1991.

Powers, M. L., "Analysis of Gravity Separation in Freewater Knockouts," *SPE Production Engineering,* 1990, pp. 52–58.

Powers, M. L., "New Perspectives on Oil and Gas Separator Performance," SPE Production and Facilities, May 1993, pp. 77–83.

Powter, C. B., "Alberta's Oil and Gas Reclamation Research Program," Proceedings of the U.S. Environmental Protection Agency's First International Symposium on Oil and Gas Exploration and Production Waste Management Practices, New Orleans, LA, Sept. 10–13, 1990, pp. 7–16.

Reid, P. I., Elliott, G. P., Minton, R. C., Chambers, B. D., and Burt, D. A., "Reduced Environmental Impact and Improved Drilling Performance with Water-Based Muds Containing Glycols," paper SPE 25989 presented at the Society of Petroleum Engineers/ Environmental Protection Agency's Exploration and Production Environmental Conference, San Antonio, TX, March 7–10, 1993.

Sarathi, P. S., "Environmental Aspects of Heavy Oil Recovery by Thermal EOR Processes," paper SPE 21768 presented at the Society of Petroleum Engineers Western Regional Meeting, Long Beach, CA, March 20–22, 1991.

Schaich, J. R., "Estimate Fugitive Emissions from Process Equipment," *Chemical Engineering Progress,* Vol. 87, No. 8, Aug. 1991, pp. 31–35.

Schumacher, J. P., Malachosky, E., Lantero, D. M., and Hampton, P. D., "Minimization and Recycling of Drilling Waste on the Alaskan North Slope," *J. Pet. Tech.,* June 1991, pp. 722–729.

Sheehan, P. E., "Air Quality Permitting of Onshore Oil and Gas Production Facilities in Santa Barbara County, California," paper SPE 21767 presented at the Society of Petroleum Engineers Western Regional Meeting, Long Beach, CA, March 20–22, 1991.

Smith, B. P., "Exposure and Risk Assessment," in *Hazardous Waste Management Engineering,* E. J. Martin and J. H. Johnson, Jr. (editors). New York: Van Nostrand Reinhold Company, Inc., 1987.

Snavely, E. S., "Radionuclides in Produced Water, A Literature Review," Report to API, Washington D.C., 1989.

St. Pe, K. M., Means, J., Milan, C., Schlenker, M., and Courtney, S., "An Assessment of Produced Water Impacts to Low-Energy, Brackish Water Systems in Southeast Louisiana: A Project Summary," Proceedings of the U.S. Environmental Protection Agency's First International Symposium on Oil and Gas Exploration and Production Waste Management Practices, New Orleans, LA, Sept. 10–13, 1990, pp. 31–42.

Stephenson, M. T., "Components of Produced Water: A Compilation of Industrial Studies," *J. Pet. Tech.,* May 1992, pp. 548–603.

Thompson, P. A., Berry, C. A., Espenscheid, A. P., Cunningham, J. A., and Evans, J. M., "Estimating Hydrocarbon Emissions from Triethylene Glycol Dehydration of Natural Gas," paper SPE 25952 presented at the Society of Petroleum Engineers/Environmental Protection Agency's Exploration and Production Environmental Conference, San Antonio, TX, March 7–10, 1993.

Thurber, N. E., "Waste Minimization in E & P Operations," Proceedings of the First International Symposium on Oil and Gas Exploration and Production Waste Management Practices, New Orleans, LA, Sept. 10–13, 1990, pp. 1039–1052.

U.S. Environmental Protection Agency, "Estimating Air Emissions from Petroleum UST Cleanups," Washington D.C., June 1989.

Wojtanowicz, A. K., "Environmental Control Potential of Drilling Engineering: An Overview of Existing Technologies," paper SPE/IADC 21954 presented at the Society of Petroleum Engineers 1991 Drilling Conference, Amsterdam, The Netherlands, March 11–14, 1991.

Wojtanowicz, A. K., Field, S. D., Krilov, Z., and Spencer, F. L., "Statistical Assessment and Sampling of Drilling-Fluid Reserve Pits," June 1989, pp. 162–170.

CHAPTER 3

The Impact of Drilling and Production Operations

Many of the materials and wastes associated with drilling and production activities have the potential to impact the environment. The potential impact depends primarily on the material, its concentration after release, and the biotic community that is exposed. Some environmental risks may be significant, while others are very low.

The most common measure of the potential environmental impact of a material is its toxicity. Toxicity occurs when a material causes a deleterious effect on an organism, population, or community. These effects can range from temporary disorientation to lethality. This chapter reviews how toxicity is measured and then summarizes many of the toxicities measured for materials associated with drilling and production activities.

3.1 MEASURING TOXICITY

The toxicity of a substance is a measure of how it impairs the life and health of living organisms following exposure to the substance. In most cases, the effects of the substance on human life and health is of primary importance. Toxicity is determined through bioassays by exposing laboratory animals to different amounts of the substance in question. The resulting effects on the health of the animals are observed. For petroleum industry wastes, common test species used for marine waters are the mysid shrimp (*Mysidopsis bahia*) and sheepshead minnow (*Cyrinidon variegatus*), while fathead minnow

(*Pimephales promelas*) and daphnid shrimp (*Ceriodaphnia dubia*) are used for fresh waters.

Two types of toxicity measurements are commonly used: *dose* and *concentration*. The dose is the concentration of a substance that has been absorbed into the tissue of the test species, while the concentration is a measure of the concentration of a substance in the environment that the species lives in. Toxicity measurements using concentration also include a time interval of exposure.

The dose is the mass fraction of the substance in the animal tissue (milligram of substance per gram of tissue, mg/g) when a particular effect has been observed. A dose that is lethal to 50% of the animals is called LD_{50}, while the lowest dose that is lethal, i.e., the dose resulting in the first death, is called LDLO. The dose levels required for any particular effect also depend on how the animal is exposed— by injection, ingestion, or inhalation.

The concentration is the fraction of the substance in air or water that causes a particular effect when the target animal is placed in that environment. It is normally given either as a mass fraction in parts per million (ppm) or as mass per unit volume in milligrams per liter (mg/l). A lethal concentration that kills 50% of the animals within a given period of time is called LC_{50}. Similarly, the lowest lethal concentration for the same period of time is called LCLO. Concentration is the toxicity measure most commonly used for materials associated with the petroleum industry.

If a material is highly toxic, then only a small concentration will be lethal and the numerical values of the lethal doses and concentrations— LD_{50}, LDLO, LC_{50}, and LCLO—would be low. Conversely, a high value of these parameters indicates low toxicity. LC_{50} values on the order of 10 are normally considered highly toxic, while values on the order of 100,000 are considered nontoxic. The length of exposure to a substance can be divided into descriptive types, as indicated in Table 3-1. Exposure that causes an immediate effect is called *acute,* while repeated, long-term exposure is called *chronic.*

There are a number of significant limitations to bioassays for toxicity testing. These limitations must be considered when new regulations are being considered or laboratory test protocols are being developed.

One limitation to most bioassay testing for toxicity is that the tests yield only acute lethal concentrations. They provide no data on the

Table 3-1
Exposure Types

Exposure Type	Duration of Exposure
Acute	Less than 24 hours
Subacute	Less than 1 month
Subchronic	1–3 months
Chronic	More than 3 months

sublethal or long-term effects of the tested substances. For sublethal effects, a related toxicity parameter, EC_{50} can also be used. This parameter is the concentration that would result in adverse effects in test species after an exposure of a specified duration. Few data are available on EC values, however, because they are difficult and expensive to quantify. A related sublethal toxicity parameter is the NOEC (no observable effect concentration), the concentration at or below which no effects are observed.

Another limitation to bioassays is that they are conducted in a laboratory and do not necessarily represent the field conditions that would actually be encountered if exposure occurred. Field conditions normally involve different concentrations and different mixtures of potentially toxic materials. Because regulations are normally based on laboratory data, these differences can lead to regulations not reflecting the actual risks in the field.

A third difficulty with laboratory bioassays is that they do not provide adequate information about chronic effects, including the mutagenic or carcinogenic activity of a substance. Bioassays normally consist of exposing the test animal to a single, high-level dose of the substance in question. Such acute exposure may not induce tumors in the test animal, even if a chronic exposure of the same substance and same total dose could. Such time-dependent responses have been observed with polyaromatic hydrocarbons.

The problem with measuring mutagenic or carcinogenic activity with bioassays is that such activity takes time to appear and be identified. Many substances, called mutagens, can alter the structure of DNA molecules in individual cells. Most mutations result in the death of the individual cells affected, with no reproduction of the mutation. If a mutated cell survives and results in future birth defects, the substance is called teratogenic. If the mutation results in cancer, the substance is called carcinogenic. As a rule, nearly all carcinogens are also mutagens, but not all mutagens are carcinogens.

For nonmutagenic or noncarcinogenic substances, a threshold dose is assumed to exist, below which there is no toxic effect. The threshold dose depends on the ability of the organism to detoxify and excrete the substance and repair any damage through normal biological processes. If an organism is exposed to a dose higher than one that can be repaired by normal biological processes, then toxic impact will occur. The magnitude of this impact will increase as the dose increases over the threshold dose. Although some substances are toxic in high concentrations, they may be essential in low concentrations for normal biological processes. These required substances include trace minerals and heavy metals commonly found in petroleum operations. Bioassays are generally not able to determine this information. For carcinogenic or mutagenic substances, however, it is assumed that there is no threshold dose. The impact is assumed to increase with the dose over all exposure levels.

Another significant limitation to bioassays is the time it takes for results to be obtained. Bioassays typically take two to three weeks to be completed. A related difficulty is that these tests are normally conducted off-site, which requires shipping of the fluid samples and delays in starting the tests. These delays can affect test results because the fluid chemistry can change over time. All of these difficulties prevent on-site decisions from being made about the fluid system, particularly drilling muds. They can result in a drilling mud from an offshore platform being shipped to shore for more expensive disposal, when it could legally be discharged overboard. A considerable effort is underway to develop more rapid bioassays for drilling fluids, particularly those that can be performed on-site.

One potentially valuable method for rapid toxicity characterization is the Microtox method (Hoskin and Strohl, 1993). In this method, a marine luminescent bacterium, *Photobacterium phosphoreum,* is used. These bacteria emit light as part of their metabolic processes. Exposure to a toxic substance interferes with these processes and results in a reduction in their light output. An advantage of the Microtox method is that the test is conducted in 15 minutes. A related process for measuring the toxicity of materials is the cumulative bioluminescence of *Pyrocystis lunula* (Wojtanowicz et al., 1992). The correlation between these two bioluminescence methods with each other and with mandated mysid shrimp toxicity assays, however, has not been good. Part of this poor correlation is from the poor reproducibility of the mysid shrimp toxicity tests between different laboratories. Mysid shrimp assays are discussed below in the section on drilling fluids.

Another rapid toxicity assay that has been studied is the fertilization rate of sea urchins. In this test, sea urchin sperm and eggs are collected, exposed to the substance being tested, and then combined. The fraction of eggs fertilized is then measured. Like the bioluminescence tests, the fertilization assays correlate poorly with the mysid shrimp toxicity tests (American Petroleum Institute, 1989f).

Another approach to toxicity testing is to develop correlations between chemical characteristics of the substance and the bioassays. One such correlation was attempted for a high-weight lignosulfonate drilling fluid (American Petroleum Institute, 1985c). The chemical characteristics studied include pH, redox potential, sulfide concentration, dissolved chromium, saturated aliphatic hydrocarbons, non-volatile aromatic hydrocarbons, straight-chain alkanes, volatile aromatic hydrocarbons, unidentified volatile hydrocarbons, and bacterial activity. Dissolved chromium was found to correlate the best with the mysid shrimp toxicity for the mud liquids. Unidentified volatile hydrocarbons correlated the best for the toxicity of the solids. These correlations, however, were poor, accounting for only about 20% of the toxic effects.

Although bioassays are conducted on animals, the results are often used to determine acceptable levels for human exposure. Animal toxicity data are extrapolated to humans, using factors like body weight and a variety of safety factors. Based on these extrapolations, a variety of human health and safety guidelines have been developed.

Human safety guidelines include *threshold limit values—time weighted average* (TLV-TWA), *threshold limit values—short-term exposure* (TLV-STEL), *threshold limit value—ceilings* (TLV-C), and *reference dose* (RfD). The TLV-TWA is the time-weighted average concentration for a normal 8-hour workday and 40-hour workweek to which nearly all workers can be chronically exposed without adverse effects. The TLV-STEL is the highest short-term exposure to which a worker can be exposed without the worker experiencing irritation, chronic or irreversible tissue change, or narcosis at a level that impairs judgment or work efficiency. The TLV-C is the concentration that should not be exceeded during the work day. TLV values are developed by the American Conference of Governmental Industrial Hygienists (ACGIH). *Reference doses* (concentration per mass of tissue) are an estimate of a daily exposure level to humans that is likely to occur without an appreciable risk of deleterious effects during a lifetime. Table 3-2 provides an example of a reference dose for several hydrocarbons.

These guidelines have been promulgated into rules and regulations by the Occupational Safety and Health Administration (OSHA). These rules are referred to as *permissible exposure levels* (PEL). TLV values are guidelines based on scientific evidence. PEL values are legal rules based on health, economic, and safety considerations. The National Institute of Occupational Safety and Health (NIOSH) also develops *recommended exposure limits* (RELs). Like TLV values, RELs are guidelines, not rules.

One source of toxicity data in the United States are Material Safety Data Sheets (MSDSs). For any substance sold in the United States, the manufacturer must provide an MSDS that summarizes all known health and physical hazard information about the substance. The toxicity information provided on MSDSs is most commonly LD_{50} data. Although the format of MSDSs can vary, they must provide the following information:

1. Manufacturer's name, address, phone number, and date of MSDS preparation.
2. Identity of material (chemical and common names).

Table 3-2
Calculated Reference Dose
for Petroleum Hydrocarbons

Hydrocarbon	Reference Dose (mg/kg-day)
Mineral spirits	0.015
Diesel fuel no. 2	0.04
Lubricating oil	0.11
Crude oil	0.04

Source: Ryer-Power et al., 1993.
Copyright SPE, with permission.

3. List of hazardous ingredients, with exposure limits.
4. Physical and chemical characteristics, including boiling point, melting point, density, solubility in water, appearance, odor, vapor density, and vapor pressure.
5. Fire and explosion hazard data, including flash point, flammability limits, extinguishing media, special firefighting procedures, and unusual fire and explosion hazards.
6. Reactivity data, including chemical stability, incompatibility with other chemicals and materials, hazards of decomposition or byproducts, and whether the material polymerizes.
7. Health hazard data, including exposure routes (inhalation, skin, or ingestion), acute and chronic health hazards, toxicity data, carcinogenicity, signs and symptoms of exposure, medical conditions aggravated by exposure, and emergency procedures (including first aid).
8. Precautions for safe handling and use, including steps to be taken if the material is spilled or released, first-aid procedures for exposure, method for disposal, and precautions for handling and storage.

One limitation to MSDS data is that it is often incomplete; normally, it only summarizes existing information from the literature. The manufacturer, in many cases, is not required to conduct additional research on the material. Such research is generally very costly and time-consuming. Because of this, the quality of MSDS data can vary considerably from chemical to chemical and from vendor to vendor. Even though manufacturers may not be required to conduct bioassays on the materials they offer for sale, bioassays are often required before a permit to discharge a material to the environment can be obtained.

3.2 HYDROCARBONS

Crude oil contains thousands of different types of hydrocarbon molecules. The toxicities and potential environmental impacts of the different molecules vary considerably. Numerous studies have been conducted on the environmental impact of hydrocarbon exposure. In this section, the major types (families) of hydrocarbons and their toxicities are discussed, and related environmental impact studies are reviewed.

3.2.1   Hydrocarbon Families

Crude oil contains thousands of different kinds of hydrocarbon molecules, making it very difficult to characterize. Crude oil can also contain significant quantities of other elements, like sulfur, nitrogen, oxygen, and heavy metals, further complicating its characterization. Crude oil is typically composed of between 50% and 98% hydrocarbons. Other important components can be sulfur (0–10%), nitrogen (0–1%) and oxygen (0–5%). Heavy metals can be found in the parts-per-million level (National Research Council, 1985).

The molecules in crude oil, however, can be grouped into a few families having similar properties. These families are distinguished primarily by how the carbon atoms bond to each other and by the presence of elements other than carbon and hydrogen. Table 3-3 summarizes most of the families of hydrocarbons found in crude oil. These families are discussed below.

Table 3-3
Families of Hydrocarbons

Family Name	Examples	Formula
Alkanes	Methane	CH_4
	Ethane	C_2H_6
	Propane	C_3H_8
Alkenes (olefins)	Methene	C_2H_4
	Propene	C_3H_6
Alkynes (acetylenes)	Ethyne	C_2H_2
	Propyne	C_3H_4
Cyclic Alkanes (naphthenes, cycloparaffins)	Cyclopropane	C_3H_6
	Cyclobutane	C_4H_8
Aromatics	Benzene	C_6H_6
	Toluene	C_6H_5CH
Polyaromatics	Naphthelene	$C_{10}H_8$
	Tetralin	$C_{10}H_{12}$
Alcohols	Methanol	CH_3OH
	Ethanol	C_2H_5OH
Acids	Acetic acid	C_2H_4OH
Amines	Methyoamine	CH_3NH_2

The simplest family consists of the alkanes. These molecules contain only carbon and hydrogen and are distinguished by the single bond between each carbon atom. All other bond sites are occupied by hydrogen. The chemical formula has the general form C_nH_{2n+2}. This family is also called paraffins or saturated hydrocarbons, because it contains the maximum possible amount of hydrogen. The chemical structure of some common alkanes is shown in Figure 3-1. Although

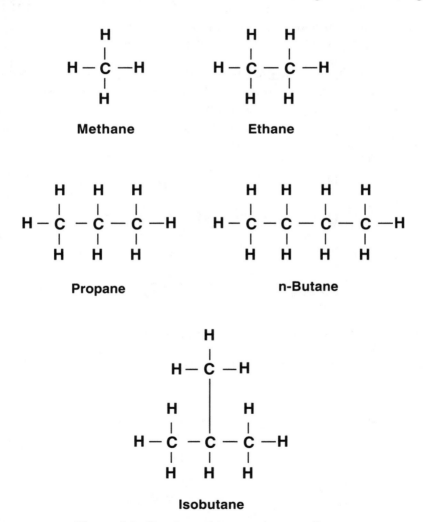

Figure 3-1. Structure of some common alkanes.

the carbon chain can branch, as seen by the two isomers of butane, the carbon chain does not form continuous loops.

The next family of hydrocarbons is the alkenes or olefins. These molecules are like the alkanes, except that one of the carbon-to-carbon bonds is a double bond instead of a single bond. For each double bond between the carbon atoms, there are two fewer bond sites available for hydrogen (one from each carbon associated with the double bond. Because of this, the chemical formula has the general form C_nH_{2n}. The chemical structure of some common alkenes is shown in Figure 3-2. Alkenes are unsaturated, because not all possible bond sites contain hydrogen.

The third family of hydrocarbons is the alkynes or acetylenes. These molecules are characterized by a triple bond between two of the carbon atoms. The resulting chemical formula has the general form C_nH_{2n-2}. The chemical structure of some common alkenes is shown in Figure 3-3. Alkynes are also unsaturated.

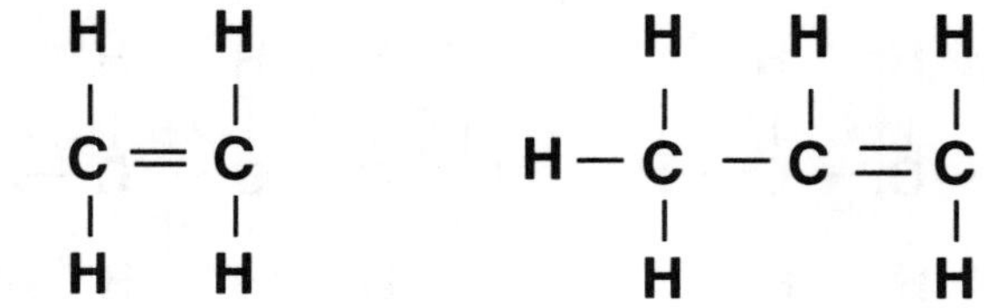

**Ethene
(Ethylene)**   **Propene
(Propylene)**

Figure 3-2. Structure of some common alkenes.

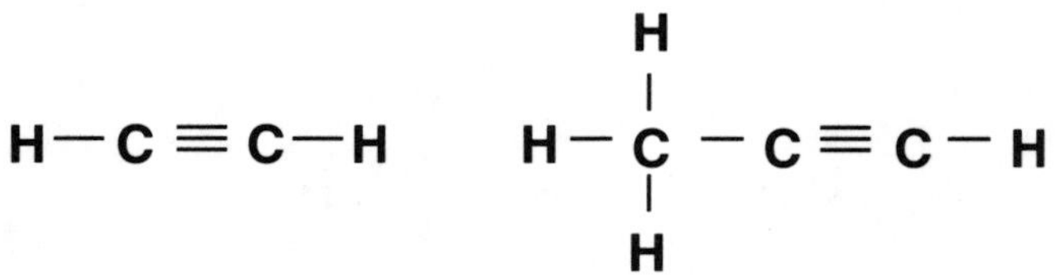

**Ethyne
(Acetylene)**   **Propyne**

Figure 3-3. Structure of some common alkynes.

The fourth family of hydrocarbons is the cyclic alkanes, also called naphthenes or cycloparaffins. These hydrocarbons have the carbon chain loop back upon itself, forming a ring or cyclic structure. All carbon-to-carbon bonds are single bonds. The chemical formula for these compounds has the general form C_nH_{2n}. The chemical structure of some common naphthenes is shown in Figure 3-4. In accordance

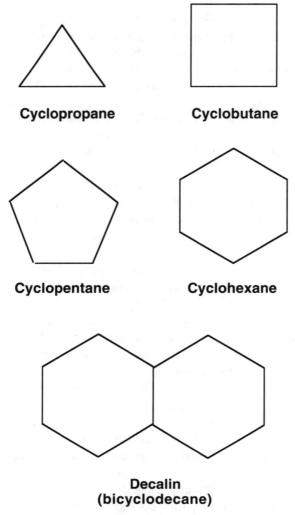

Figure 3-4. Structure of some common cyclic alkanes.

with common organic chemistry symbolism used for more complex molecules, only the carbon-to-carbon bonds are shown in this figure. The carbon atoms are found at the intersections of the bonds, and the hydrogen atoms are inferred around the carbon atoms such that the four carbon bond sites are all occupied. For naphthenes, two hydrogen atoms are found with each carbon atom. Some hydrocarbons can have multiple rings, with shared carbon atoms. An example of one of these condensed rings is decaline.

The fifth family of hydrocarbons is the aromatics. These compounds are also ring structures, but each carbon has only one hydrogen atom and the remaining bond sites are shared among the adjacent carbon atoms. This results in very stable carbon-to-carbon bonds. These bonds are conveniently written as an alternating double-single bond, as shown in Figure 3-5, although each carbon-to-carbon bond is identical. Benzene is the simplest of the aromatic hydrocarbons. Other aromatics can be created by replacing one of the hydrogen atoms with a carbon chain, as shown in Figure 3-5. Three isomers of xylene are also possible, with only one isomer shown in the figure.

A sixth family of hydrocarbons is the polyaromatic hydrocarbons. Condensed aromatics are also known as polycyclic aromatics, polyaromatic hydrocarbons (PAH), or polynuclear aromatics (PNA). These condensed ring structures have aromatic rings sharing carbon atoms with other rings. Two examples are shown in Figure 3-6. The polyaromatic fraction of crude oil ranges between about 0.2% and 7.4%.

Other families of hydrocarbons contain atoms other than carbon and hydrogen. Alcohols are formed by replacing a hydrogen atom with an oxygen-hydrogen atom pair. Organic acids are formed by replacing the three hydrogen atoms at the end of a hydrocarbon chain with a double bonded oxygen atom and an oxygen-hydrogen atom pair. Amines are formed by replacing a hydrogen atom with a nitrogen atom having two hydrogen atoms bonded to it. The chemical structures of several such compounds are shown in Figure 3-7.

Other families of hydrocarbons can be created if a carbon atom in the carbon chain or ring is replaced by other elements. Sulfur and nitrogen are commonly found as a carbon substitute. Heavy metals are found in complex compounds called porphyrins.

A final family of hydrocarbons is the asphaltenes. These are large polyaromatic hydrocarbons that contain sulfur, oxygen, or nitrogen. They contain typically three to ten ring structures. Pure asphaltenes

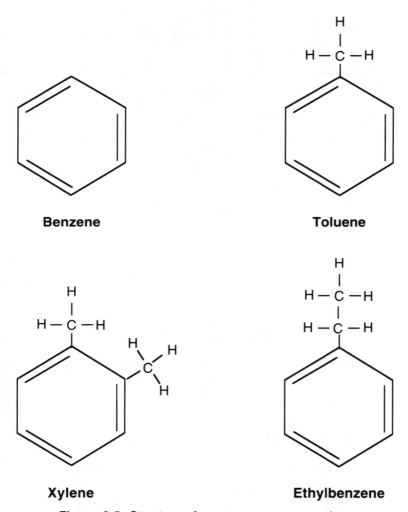

Figure 3-5. Structure of some common aromatics.

are solids and are insoluble in crude oil, although they can be dispersed in oil as a colloidal suspension.

3.2.2   Hydrocarbon Toxicity

A number of bioassay tests have been conducted to determine the toxicity of various hydrocarbons on marine animals. The toxicity of

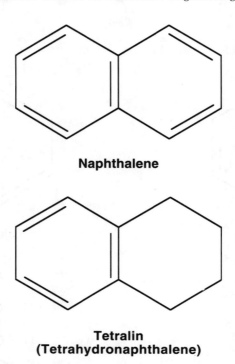

Naphthalene

Tetralin
(Tetrahydronaphthalene)

Figure 3-6. Structure of some common polyaromatic hydrocarbons.

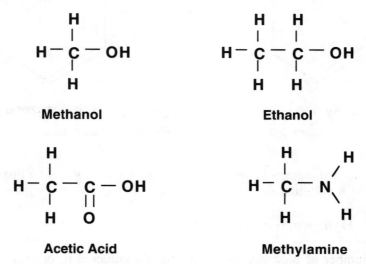

Methanol

Ethanol

Acetic Acid

Methylamine

Figure 3-7. Structure of mixed hydrocarbon compounds.

hydrocarbons has been found to vary considerably and generalizations cannot be easily made. Factors that affect toxicity include molecular weight, hydrocarbon family, the organism exposed to the hydrocarbon, and life-cycle stage of the organism exposed (egg, larva, juvenile, or adult). For mixtures of hydrocarbons, such as crude oil, the toxicity also depends on the history of the exposure.

For hydrocarbons of a similar type (the same family), the toxicity tends to increase with decreasing molecular weight. Smaller molecules tend to be more toxic than large molecules. Light crude oils and refined products tend to be more toxic than those of heavy crude oils, because heavy crude oils have a higher average molecular weight. For similar molecular weight hydrocarbons, the toxicity varies with family. The toxicity of hydrocarbon families generally increases in the following order: alkanes, alkenes, cycloparaffins, aromatics, and polyaromatic hydrocarbons.

Some of the least toxic hydrocarbons include dodecane and higher paraffins. In fact, these high molecular weight paraffins are used in cooking, food preparation, and candles. The most toxic hydrocarbons are the low-boiling-point aromatics, particularly benzene, toluene, ethylbenzene, and xylene. Because of their similar properties, these four aromatic molecules are commonly referred to as BTEX. The most toxic hydrocarbons also tend to have a high solubility in water. A high solubility makes a molecule more accessible for uptake by plants and animals.

The toxicity of a given hydrocarbon varies considerably with the organism exposed. Factors that also affect the toxicity to a particular organism include the general health of the organism and whether the organism is already stressed. Stress factors include water salinity, temperature, and food abundance. The toxicity of crude oil to some fish can be twice as high in seawater as in fresh water. The toxicity of a particular hydrocarbon also appears to increase with decreasing temperature. Synergistic effects from the presence of other toxins can also significantly alter the toxicity of specific hydrocarbons.

The toxicities (LC_{50}) for a variety of aromatic and polyaromatic hydrocarbons are shown in Tables 3-4a and 3-4b (National Research Council, 1985). The LC_{50} values for many aromatic hydrocarbons are less than about 5 ppm, although some have values as high as 28 ppm. From these tables, it can also be seen that the toxicity is higher (lower LC_{50}) for higher molecular weight polyaromatic

Table 3-4a
Summary of Bioassay Tests on Marine Organisms

Hydrocarbon	Test Species	Test Duration (hr)	LC_{50} (ppm)
Benzene	Grass Shrimp	96	27
	Crago	96	20
	Striped Bass	96	6
Toluene	Grass Shrimp	96	9.5
	Crago	96	4
	Striped Bass	96	7.5
	Cancer	96	28
	Salmon Fry	24	5.5
m-Xylene	*Palaemonetes*	96	3.5
	Striped Bass	96	9
	Cancer	96	12
o-Xylene	*Crago*	96	1
	Striped Bass	96	11
	Cancer	96	6
p-Xylene	*Crago*	96	2
	Striped Bass	96	2
Ethylbenzene	*Palaemonetes*	96	0.5
	Striped Bass	96	5
	Cancer	96	13
	Copepod	24	3.5
Trimethylbenzene	*Cancer*	96	2
Naphthalene	*Palaemonetes*	96	2.5
	Salmon Fry	24	1
	Amphipod	96	2.5
	Neanthes	96	3.5
	Paneaus aztecus	24	2.5
	Cyprinodon	24	2.5
Methylnaphthalene	*Palaemonetes*	96	1
	Cancer	96	2
	Copepod	96	1.5
	Penaeus aztecus	24	0.5
	Cyprinodon	24	3.5
	Copepod	24	2

Source: after National Research Council, 1985.
Copyright © 1985, National Academy of Sciences.
Courtesy of National Academy Press, Washington, D.C.

Table 3-4b
Summary of Bioassay Tests on Marine Organisms

Hydrocarbon	Test Species	Test Duration (hr)	LC_{50} (ppm)
Dimethylnaaphthalene	*Palaemonetes*	96	0.5
	Cancer	96	0.5
	Neanthes	96	2
	Penaus aztecus	24	0.5
	Copepod	24	0.5
	Cyprinodon	24	5
Trimethylnaphthalene	*Neanthes*	96	2
	Cancer	24	0.25
Fluorene	*Palaemonetes*	96	0.25
	Neanthes	96	1
	Cyprinodon	96	1.5
Dibenzothiophene	*Palaemonetes*	96	0.25
	Cyprinodon	96	3
Phenanthrene	*Palaemonetes*	24	0.25
	Neanthes	96	0.5
Methylphenanthrene	*Neanthes*	96	0.25
Fluoranthene	*Neanthes*	96	0.5

Source: after National Research Council, 1985.
Copyright © 1985, National Academy of Sciences.
Courtesy of National Academy Press, Washington, D.C.

hydrocarbons than the single ring aromatics of benzene, toluene, ethylbenzene, and xylene (BTEX).

The high toxicity of aromatic hydrocarbons relative to other hydrocarbons can be seen by comparing the 96-hour mysid shrimp toxicity for drilling muds using diesel oil to that using mineral oils. Diesel oil contains as much as 60% aromatic components, while some mineral oils contain less than 1%. LC_{50} values for diesel are around 2,000 ppm, while those for some mineral oils are greater than 1,000,000 ppm, in which case less than 50% of the test species died during the test period (Derkics and Souders, 1993). As discussed below in the section on drilling fluids, the mysid shrimp test protocol dilutes the oil with sea water by a factor of nine before the test is conducted. Thus, these mysid shrimp data cannot be directly compared to data that were obtained using a protocol that does not require the same dilution.

The presence of mineral oil-based mud and synthetic oil-based mud (polyalphaolefin) on cuttings at concentrations up to 8.4% had no significant effect on the growth of mud minnows (*Fundulus grandis*). The uptake of mineral oils, however, was higher than that of synthetic oils, suggesting that synthetic, high-molecular weight liquids may have a lower toxicity (Rushing et al., 1991; Jones et al., 1991).

For a particular organism, the life-cycle stage at which exposure occurs can impact how toxic a material is. Table 3-5 shows the results of bioassays on four organisms as a function of life-cycle stage for exposure to No. 2 fuel oil. From this table, it can be seen that some species have a higher tolerance at younger stages, while other species have a higher tolerance at older stages (National Research Council, 1985). In most species, however, the adults are more tolerant of

Table 3-5
Effect of Life-Cycle Stage on Fuel Oil Toxicity

Species	96-hour LC_{50} (ppm)
Brown Shrimp:	
Postlarvae	6.6
Small juveniles	3.8
Large juveniles	2.9
White Shrimp:	
Postlarvae	1.3
Juveniles	1.0
Grass Shrimp:	
Larvae	1.2
Postlarvae	2.3
Adults	3.6
Polychaeta:	
4 segments	8.3
18 segments	5.8
32 segments	5.5
Adults (40 segments)	4.0

Source: after National Research Council, 1985.
Copyright © 1985, National Academy of Sciences.
Courtesy of National Academy Press, Washington, D.C.

exposure to hydrocarbons than the young. For all of the species included in this table, however, the LC_{50} values are below 10 ppm, indicating a high toxicity at all life-cycle stages.

An important factor affecting the toxicity of crude oils is their history before any organisms are exposed. Because the most toxic hydrocarbons are also the most volatile, they rapidly evaporate from a release site. Within a few days after a crude oil release, only higher molecular weight hydrocarbons remain, so the toxicity of the remaining crude oil is lower. Hydrocarbons in water also tend to adsorb onto suspended sediments, making them much less bioavailable to marine organisms than hydrocarbons in solution or dispersion in water. This further lowers the toxicity of released crude oil. If the sediments accumulate on the bottom of the sea, they can accumulate in estuarine organisms. The accumulation and metabolism of these compounds, however, vary with species (American Petroleum Institute, 1989e).

Impact of Crude Oil on Marine Animals

The actual impact of hydrocarbon exposure on marine animals is more complex than simple bioassay tests reveal. Oil at sublethal concentrations can significantly alter the behavior and development of marine organisms. These effects, however, are difficult to quantify. The problem of determining sublethal toxicity is further compounded because different species have different reactions and there is mixed effect when multiple toxins are present. Although there is a tremendous amount of scatter in the data, most threshold concentrations of crude oil in water for effects to be observed for eggs, embryos, and larvae of marine fish are between 0.01 and 5 mg/l (National Research Council, 1985).

Behavioral changes from exposure to hydrocarbons are primarily those involving motility, while in higher organisms, changes affect avoidance, burrowing, feeding, and reproductive activities. Behavioral changes in feeding have been observed at hydrocarbon concentrations as low as a few microgm/l. Other measures of sublethal effects include changes in respiration, the ratio of oxygen consumed to nitrogen excreted, biochemical enzyme assays, and cellular activity. The respiratory rate following exposure is usually reduced, although in some cases, it is increased. The level of exposure for respiratory impact for fish and planktonic crustaceans in the laboratory is less than 1 mg/l.

Continued hydrocarbon exposure also lowers the growth rate of animals (National Research Council, 1985).

Exposure to hydrocarbons can adversely affect the development of organisms in some species at concentrations below 1 mg/l. Some species (annelids, gastropods, and copepods) show no long-lasting damage, while other species (corals, bivalves, and some crustaceans) can suffer long-term damage at an oiled site (National Research Council, 1985).

The impact of hydrocarbon exposure also depends on whether the hydrocarbon is dissolved or dispersed as suspended droplets. For shrimp, the toxicity of dispersed crude oil was found to decrease with decreasing amounts of total aromatic hydrocarbons (benzene, alkyl-benzenes, and naphthalenes). For sand lance fish, however, the impact could not be tied directly to the concentration of aromatics in the water. Instead, it was postulated that the oil droplets attached to their eggs and cut off their oxygen supply (American Petroleum Institute, 1985a).

One concern with crude oil spills is their potential impact on the behavior of migratory salmon. Because salmon identify their home water by smell, there was concern over whether their sense of smell would be affected by passing beneath a spill so that they could not recognize their home water. Studies have shown that as long as the fish pass back into uncontaminated water, their homing ability is not affected (American Petroleum Institute, 1985b; American Petroleum Institute, 1986a; American Petroleum Institute, 1987a).

The effects of spilled crude oil on the reproductive success of Pacific herring (*Culpea harengus pallasi*) have also been studied (American Petroleum Institute, 1985a). No effect in fertilization rates or total percentage of eggs successfully hatched were observed. However, exposure to oil significantly increased the frequency of abnormal larvae. These abnormalities included spinal deformities, swollen pericardial regions, and yolk sac compartmentalization. The amount of oil droplets adhering to the eggs apparently controlled the frequency of abnormal larvae rather than the total oil concentration in the water. It was not certain whether this increase of abnormal larvae resulted from the toxic compounds of the oil passing to the eggs or from oxygen deprivation from the oil droplets covering the egg.

The most common impact of crude oil on birds is by direct contact. Oil coats their feathers, causing them to lose their water-repellance and thermal insulation. The birds then sink and drown or die of hypothermia. Oil can also be ingested by the birds during preening of oiled plumage. Although this oil becomes distributed throughout the body, there is no evidence that ingested oil is a primary cause of death among birds (National Research Council, 1985).

Nonlethal exposures of birds to crude oil significantly reduces hatchling success and fledgling success in a dose-dependent manner (American Petroleum Institute, 1988). Adult petrels were most sensitive to contaminant exposure late in the incubation period and early in the post-hatchling period. Pollutant-related decreases in reproductive success were probably associated with temporary abandonment of the nesting burrow by adults. Treated adults generally returned to normal behavior in the second season following exposure. The primary concern for marine birds appears to be the immediate effects on adult mortality and the resulting population dynamics. The effects of sublethal exposures may be significant only in areas where exposure is chronic.

Crude oil also impacts phytoplankton and zooplankton. The effect of crude oil is to inhibit growth and photosynthesis (for phytoplankton) at concentrations in the range of 1–10 mg/l (National Research Council, 1985).

The effect of oil on marine mammals is highly variable. Fur-insulated mammals lose their ability to thermally regulate their temperature as their oil-contaminated fur loses its insulating capacity. The loss of thermal insulation creates a higher metabolic activity to regulate body temperature, which results in fat and muscular energy reserves being rapidly exhausted. This can result in the animal's death by hypothermia or drowning. Many species show no avoidance response to oiled areas. Chronic contact of marine mammals with oil may also result in skin and eye lesions (National Research Council, 1985).

Impact of Crude Oil on Ecosystems

Only a few studies have been conducted on the chronic effects of hydrocarbon releases on ecosystems. No apparent long-term impacts on the productivity of ecosystems have been observed. In all cases, the affected areas recovered after the hydrocarbon source had been removed, although full recovery could take a number of years. One difficulty with ecosystem studies, however, is that little is known about ecosystems that have not been exposed to hydrocarbons. This makes it difficult to determine what lasting effects hydrocarbons do have on ecosystems (National Research Council, 1985).

One ecosystem that is chronically exposed to hydrocarbons from petroleum production is the Gulf of Mexico. Natural variations in this ecosystem, however, may completely mask any effects of oil production (National Research Council, 1985). Natural variations in the ecosystem cause large changes in the diversity and number of organisms present at any particular location. These natural variations include the effects of the Mississippi River discharging into the Gulf. The Mississippi River has a low salinity, a low oxygen concentration, a high concentration of nitrogen fertilizers, and a high concentration of suspended solids. These conditions vary significantly from those of the marine environment and can have a significant natural impact on the ecosystem. Because of the chaotic nature of the turbulent discharge of the Mississippi River, this impact occurs over a large portion of the Gulf.

A study of hydrocarbon and heavy metal contamination on the continental shelf of the Louisiana coast in the Gulf of Mexico revealed contamination of hydrocarbons and heavy metals near offshore platforms (U.S. Bureau of Land Management, 1981). No measurable impact on the ecosystem could be observed from the presence of the offshore platforms. In areas having a very low background level of hydrocarbons in the sediments, elevated hydrocarbon levels were observed at distances up to 2,000 m from the platform. In areas having a relatively high background level of hydrocarbons, no concentration of hydrocarbons in the sediments around the platform was observed. All levels were, however, below those for public concern. Essentially no accumulation of hydrocarbons in organisms around platforms was observed. Some concentration of heavy metals occurred, but insufficient data were obtained for a reliable statistical analysis. The Mississippi River, with its fresh water, high sediment load, and low dissolved oxygen content, had a greater measured impact on the benthic ecosystem than did the offshore platforms.

Contamination of sediments by polyaromatic hydrocarbons from routine discharges of produced water into shallow estuaries have been reported as far away as one kilometer (Rabalais et al., 1990). The effect of hydrocarbon contamination on the benthic community around such discharge points was correlated to the hydrocarbon contamination level. Macrobenthic fauna were missing or greatly affected when the polyaromatic hydrocarbon concentrations were on the order of 1 ppm. Reduced fauna concentrations were observed for polyaromatic hydrocarbon concentrations on the order of 0.1 ppm.

A four-year study was conducted on the ecological effects, chemical fate, and microbial responses of marsh systems following crude oil spills (American Petroleum Institute, 1981a). The effects of oil spills on phytoplankton were short lived, with a recovery within seven days to the levels found in the control area. The effects of oil on standing marsh plants were severe during the first year following exposure. Growth during the third year was, on the average, as great as in the control area, although growth in the high plant concentration areas was still lower.

The impact of chronic exposure to tar balls on intertidal biota in a rocky shore community in Bermuda have also been studied (American Petroleum Institute, 1984a). There was no correlation between the presence or amount of tar on the shore and the reproductive status of the six intertidal species studied. Snail size was correlated with the presence of tar, however. Little accumulation of hydrocarbons in tissues of intertidal animals was found. Tar balls are believed to come from discharged tanker ballast tanks, with the level of tar on a beach being controlled by the amount of direct exposure to constant wave action, topography, and configuration of the shoreline. Tar balls are accumulated almost exclusively in the upper intertidal and splash zone.

One important way to gain information about the effects of chronic exposure of ecosystems to crude oil is to study areas having natural oil seeps. Studies at natural seeps at Coal Oil Point in the Santa Barbara Channel, California, have shown that the level of macrofauna is reduced when the hydrocarbon content in the sediments is high (National Research Council, 1985; American Petroleum Institute, 1980). The reason for the lower faunal level is the reduced amount of oxygen, high sulfide content, and high level of dissolved hydrocarbons (mostly aromatics) in the surrounding water. Aromatic concentrations in water have been measured as high as 1.3 mg/l. Areas with lower seepage (less than 0.1 mg/l) show little or no impact.

A separate study of a major natural seep area near Santa Barbara, California, that leaks 50–70 barrels of oil per day revealed little impact. The growth rate of resident marine organisms near the seeps was not affected, the total biomass (plant and animal life) and biomass of individual species groups were not related to the presence of hydrocarbons in sediments, and all species expected to be in the area were indeed present. Fish feeding around the seeps did show high levels of enzyme activity needed to break down and digest the toxic oil compounds (American Petroleum Institute, 1984b).

Impact on Human Health

The impact of hydrocarbons on human health depends somewhat on whether exposure was from ingestion, inhalation, or dermal (skin) contact and on whether the exposure was acute (short-term) or chronic (long-term).

The acute effects of ingestion may include irritation to the mouth, throat, and stomach, and digestive disorders and/or damage. Small amounts of hydrocarbons can be drawn into the lungs, either from swallowing or vomiting, and may cause respiratory impact such as pulmonary edema or bronchopneumonia.

The chronic effects of ingestion may include kidney, liver, or gastointestinal tract damage, or abnormal heart rhythms. Prolonged and/or repeated exposure to aromatics like benzene may cause damage to the blood-producing system and serious blood disorders, including leukemia. The metabolism of aromatic hydrocarbons after ingestion can result in the creation of mutagenic or carcinogenic derivatives, even if the original hydrocarbon is relatively nontoxic. (National Research Council, 1985). A number of PAHs have been linked to cancer of the skin, lung, and other sites on the body. There is no epidemiologic evidence for human cancer from intake of PAH-contaminated food, however. Most human exposure to PAHs comes from nonpetroleum sources, including cigarette smoke, fossil fuel combustion products, and food.

The acute symptoms of hydrocarbon exposure by inhalation may include irritation of the nose, throat, and lungs, headaches and dizziness, anesthetic effects, and other central nervous system depression effects. These symptoms can occur at air concentrations of 0.5 mg/l for 30 minutes (Hastings et al., 1984). Epileptic-type seizures may occur months after a high acute exposure to gasoline vapors, and permanent brain damage has been reported. Acute toxic effects are not commonly observed, however, in gas station attendants and auto mechanics.

Chronic effects of inhalation exposure to hydrocarbons containing high concentrations of aromatic compounds, including gasoline, can be weight loss from loss of appetite, muscular weakness and cramps, sporadic electroencephalography irregularities, and possible liver and renal damage.

Exposure of eyes and skin to hydrocarbons may result in irritation, mechanical or chemical damage to eye tissue, or dermatitis. Long-term exposure to vacuum distillates has caused skin cancer in animals. Exposure to petrochemicals, particularly polyaromatic hydrocarbons, increases susceptibility to skin infections, including skin cancer when there is simultaneous exposure to sunlight (Burnham and Bey, 1991; Burnham and Rahman, 1992).

One potential source of hydrocarbon exposure to humans is ingestion of hydrocarbon-contaminated food, particularly seafood. Studies have shown that most organisms cleanse themselves of hydrocarbons within a matter of weeks after being removed from the source of contamination. This cleansing time, however, depends upon the contaminated organism.

The exposure levels of humans to polyaromatic hydrocarbons from crude oil may be lower than those from other, more common sources like grilled food and combustion products, or from naturally occurring sources like coffee, grains, and vegetables (American Petroleum Institute, 1978).

Suggested standards for human exposure to petroleum hydrocarbons varies with the specific hydrocarbon, but ranges between 25 and 430 ppm (National Research Council, 1985). Permitted occupational exposure levels to benzene are on the order 10 ppm, but vary with the prevailing regulations.

Impact on Plant Growth

Hydrocarbons also impact plant growth when released on land. Levels of oil and grease above a few percent in soils (by weight) have shown degradation of plant growth. Levels below a few percent have shown an actual enhancement of some crop growth. Recovery of an exposed site after a one-time hydrocarbon release usually occurs after a few months (Deuel, 1990). A level of 1% oil and grease is recommended as a practical threshold where the hydrocarbons become detrimental to plant life (American Petroleum Institute, 1989b).

Airborne hydrocarbons emitted during blowouts can also impact plant growth around the wellhead. Long-term growth rate reductions have been observed in coniferous forest growth following blowouts at distances as great as 2 km from the wellhead (Baker, 1991).

3.3 SALT

Salt (sodium chloride) in low concentrations is essential to the health of plants and animals. At concentrations different from the naturally occurring levels found in a given ecosystem, however, salt can cause an adverse impact.

3.3.1 Impact on Plants

The impact of salt on plants arises primarily from an excess salt concentration in the cellular fluids of the plants or from an alteration in the soil structure in which the plants grow. The primary impact of an abnormal salt concentration in cellular fluids is the disruption of the fluid chemistry balance within cells. This disruption inhibits cellular growth, water uptake, and the overall health of the plants. Growth of nonmarine plants is impaired at total dissolved salt concentrations between about 1,500 and 2,500 mg/l, although this threshold level varies significantly with plant type, how the water is applied, and whether the soil is kept saturated. Salt concentrations below about 1,000 mg/l seem to improve some plant growth (Vickers, 1990).

When salt was spread over soil in the form of salty drilling muds, the yield of brome grass was reduced when the concentration of chloride exceeded about 1,000 kg Cl/hectare for potassium and sodium chloride, and about 50 kg Cl/hectare for a freshwater gel. The plant yield for intermediate chloride application levels was higher than that of control plots (Macyk et al., 1990).

Salt can indirectly impact plant growth by altering the physical properties of soil. When saline water is discharged on land, it can alter the pore structure of the soil by causing compaction, limiting the access of air and water to the plant roots. The impact varies, however, with salinity level and plant type. If the total dissolved solids content is above about 2,800 mg/l, salt can build up in the soil (Vickers, 1990).

Excess sodium in soil can also cause clays to disperse, lowering the permeability of the soil. This can form an impenetrable surface crust that hinders the emergence of seedlings and limits the availability of nutrients such as iron, manganese, calcium, and magnesium to the plants (Kaszuba and Buys, 1993). On the other hand, the addition of clays from drilling muds can increase the water holding capacity of sandy/coarse-textured soils, improving plant growth.

A number of ways to measure the salinity of soils has been developed. These measurements include directly measuring the electrical conductivity of the soil and various measurements of sodium concentration.

The *electrical conductivity* (EC) of a solution is a measure of the total amount of cations and anions dissolved in water. These ions can include sodium (Na), calcium (Ca), magnesium (Mg), potassium (K), chloride (Cl), sulfate (SO_4), bicarbonate (HCO_3), carbonate (CO_3), and hydroxide (OH). The electrical conductivity is the reciprocal resistance of the solution. Table 3-6 summarizes the effects of different EC values on crops (U.S. Salinity Staff, 1954). A level of salinity that will not adversely impact most vegetation, land, or groundwater resources from the one-time discharge is one at which the electrical conductivity of the discharged brine is less than 4 mmho/cm. This level will limit the reduction of crop yields to less than 15% (Deuel, 1990).

The electrical conductivity is related to the *total dissolved solids* (TDS) concentration in the water. The TDS is the weight of residue after all of the water has been evaporated. The TDS has units of mass/volume of solution. The relationship between EC and TDS is given as follows:

$$TDS = A*EC \qquad (3-1)$$

where **A** is an empirical constant equal to about 640 (Tchobanoglous and Burton, 1991). The units of the constant are cm-mg/mmho/liter.

Table 3-6
Effect of Electrical Conductivity (EC) on Crops

EC Range (mmhos/cm)	Effect
0–2	Negligible
2–4	Yield of very sensitive crops impacted
4–8	Yield of many crops impacted
8–16	Only tolerant crops still produce
>16	Only very few tolerant crops still produce

Source: U.S. Salinity Staff, 1954.

The most common impact of brine on plants is that it increases the osmotic pressure of the soil solution. Osmosis is a process that controls the movement of water between solutions, with water flowing from lower to higher osmotic pressure. Plants have an osmotic pressure in their cells, which varies from species to species. If the osmotic pressure in the soil solution outside the plant exceeds that inside the cell, water cannot flow into the plant. High osmotic pressure produced by soluble salts also retards water imbibition by seeds, resulting in decreased germination and slower seedling emergence rates, and disrupts the uptake of nutrients in plants.

The osmotic pressure (OP) is related to the EC through the following equation (Deuel, 1990):

$$OP = 0.36*EC \qquad (3-2)$$

In this equation, the osmotic pressure is in atmospheres and the electrical conductivity is in mmho/cm.

The capacity of a soil to adsorb positively charged ions (cations) is called the *cation exchange capacity* (CEC). The exchangeable cations in a soil are those held on surface exchange sites and are in equilibrium with the soil solution. The measure of the degree that the exchange sites are saturated with sodium is called the exchangeable sodium percentage (ESP) and is calculated through the following equation:

$$ESP(\%) = \frac{NaX}{CEC} * 100 \qquad (3-3)$$

where **NaX** is the amount of exchangeable sodium. Both the CEC and NaX are expressed in units of meq/100 g. In fertile soils, the most common exchangeable cations are calcium and magnesium. These ions are less soluble than sodium and do not affect plant growth to the same degree.

For ESP greater than 15%, some soils can lose their structure and disperse in water. Dispersive soils are devastating to plant life because they limit the free exchange of air and infiltration of water (American Petroleum Institute, 1989b).

The *sodium adsorption ratio* (SAR) is an empirical mathematical expression used to characterize the detrimental effects of sodium on soils. It is calculated through the following equation:

$$SAR = \frac{Na^+}{\sqrt{\dfrac{Ca^{2+} + Mg^{2+}}{2}}} \tag{3-4}$$

where the cation concentrations are expressed in millimoles/liter. Concentrations are determined by direct chemical analysis of reserves pit liquids or aqueous extracts of waste solids or soils. High sodium levels (SAR greater than 12) in soil solutions cause Ca and Mg deficiencies in plants (American Petroleum Institute, 1989b).

3.3.2 Impact on Aquatic Organisms

Most, but not all, produced waters have a salt content higher than that found in the local ecosystems. The discharge of water having a higher salt content can impact aquatic organisms. High concentrations of sodium chloride can affect the development of embryos and fetuses and can cause fetal death. High salt concentrations can also affect the development of the musculoskeletal system and cause eye, skin, and upper respiratory system irritation.

Bioassay tests have been conducted with brines to determine the toxicity of various salts to aquatic organisms. Common freshwater species used for these tests include the water flea, rainbow trout, and the fathead minnow. As seen in Table 3-7, 48-hr LC_{50} values for the

Table 3-7
Toxicity of Salts to Water Flea, 48-hr LC_{50} (mg/L)

Salt	Anion	Cation	Total
KCl	270	290	560
K_2SO_4	400	330	730
$KHCO_3$	300	200	500
NaCl	1,300	840	2,140
Na_2SO_4	2,500	1,260	3,760
$NaHCO_3$	740	260	1,000
$CaCl_2$	1,200	700	1,900
$CaSO_4$	1,430	600	2,030
$MgCl_2$	730	250	980
$MgSO_4$	1,400	360	1,760

Source: after Mount et al., 1993.
Copyright SPE, with permission.

water flea for a variety of pure salts are on the order of 1,000 mg/l (Mount et al., 1993). Studies indicate that a concentration of 230 mg/l for total dissolved solids may be sufficient to protect warmwater species in natural streams. No significant change in macro invertebrate behavior was observed below a level of 565 mg/l (Vickers, 1990).

Because the salinity of many produced waters is greater than that of marine waters, the environmental impact of high salt concentrations is also of concern regarding marine organisms. Highly saline water has a higher density than seawater and will segregate to the bottom of any surface waters. This density gradient inhibits the mixing and dilution of the very salty water. This segregation is only a problem in shallow estuaries and marshes that allow little dilution (St. Pe et al., 1990).

The impact of a saline brine spill in a saltwater marsh was observed in 1989 following a spill of about 35 million gallons of brine (Bozzo et al., 1990). The salinity of the brine varied between 0 and 274 parts per thousand (ppt), with 17 million gallons having a salinity over 220 ppt. In comparison, seawater has a salinity of 35 ppt. Following the spill, vegetation in areas with poor drainage and along drainage channels was completely killed. Flushing from rainwater, turbulent mixing from nearby barge traffic, and tidal events lowered the salinity in the soils around the spill to ambient levels within a few months. Salt-tolerant plants began growing in the areas where the salt had killed the less tolerant plants. The following year, vegetation in all areas except those most severely affected showed signs of recovery.

3.4 HEAVY METALS

The heavy metals encountered in drilling and production activities are related to a variety of environmental concerns, depending on the metal and its concentration. At very low concentrations, some metals are essential to healthy cellular activity. Essential metals include chromium, cobalt, copper, iodine, iron, manganese, molybdenum, nickel, selenium, silicon, vanadium, and zinc (Valkovic, 1978). At high concentrations, however, metals can be toxic. Because most concentrations encountered during drilling and production are relatively low, the environmental impact is generally observed only after chronic exposure.

The environmental impact of heavy metals is manifested primarily through their interaction with enzymes in animal cells. Enzymes are complex proteins that catalyze specific biochemical reactions. Heavy metals affect the action of enzymes. Excess concentrations of metals inhibit normal biochemical processes in cells. This inhibition can result in damage to the liver, kidney, or reproductive, blood forming, or nervous systems. These effects may also include mutations or tumors. Many metals can impact embryo and larval states of fish and benthic invertebrates.

The toxicities of many metals found in the upstream petroleum industry have been summarized by the American Conference of Governmental Industrial Hygienists (ACGIH) and are listed in Table 3-8 (Proctor et al., 1989). This table lists the *threshold limit values* (TLV) for airborne exposures.

The toxicity of trace metals in agricultural soils is summarized by Logan and Traina (1993) and is given in Table 3-9. This table identifies whether the element is essential, beneficial, or toxic to plants and animals. Also found is a typical concentration of each metal in soils. From this table it can be seen that many metals are essential in low concentrations, but toxic in high concentrations.

A description of the health impacts of a number of heavy metals is given below. Further information about these and other metals is available in the literature, for example, Valkovic (1978), Proctor et al.

Table 3-8
Concentration Limits for Heavy Metals

Metal	TLV (mg/m$_3$)
Aluminum	2.0
Arsenic	0.2
Barium (soluble compounds)	0.5
Barium (barium sulfate)	10
Cadmium	0.05
Chromium (trivalent)	0.5
Chromium (hexavalent)	0.05
Lead	0.15
Mercury	0.05
Nickel (soluble inorganic compounds)	0.1
Vanadium (as vanadium pentoxide)	0.05
Zinc (as zinc oxide)	5

Source: Proctor et al., 1989.
Copyright Van Nostrand Reinhold, with permission.

Table 3-9
Role of Trace Metals in Plants and Animals

Metal	Essential to Plants	Beneficial to Animals	Toxic to Plants	Toxic to Animals	Typical Concentration (mg/kg)
Antimony	No	No	?	Yes	1.5
Arsenic	No	Yes	Yes	Yes	7
Barium	No	Possible	Low	Low	500
Beryllium	No.	No	Yes	Yes	2
Bismuth	No	No	Yes	Yes	0.2
Boron	Yes	No	Yes	—	20
Cadmium	No	No	Yes	Yes	0.35
Chromium	No	Yes	Yes	Yes (Cr^{6+})	75
Cobalt	Yes	Yes	Low	Low	9
Copper	Yes	Yes	Yes	Yes	22
Lead	No	No	Yes	Yes	25
Manganese	Yes	Yes	Yes	Low	700
Mercury	No	No	No	Yes	0.07
Molybdenum	Yes	Yes	Yes	Yes	1.5
Nickel	Possible	Yes	Yes	Yes	30
Selenium	Yes	Yes	Yes	Yes	0.3
Silver	No	No	No	Yes	0.05
Tin	No	Yes	?	Yes	4
Tungsten	No	No	?	?	1.5
Vanadium	Yes	Yes	Yes	Yes	75
Zinc	Yes	Yes	Yes	Yes	60

Source: adapted from Logan and Traina, 1993.
Reprinted from Metals in Groundwater, *with permission. Copyright Lewis Publishers, an imprint of CRC Press, Boca Raton, Florida.*

(1989), Calabrese and Kenyon (1991), St. Pe et al. (1990), and the American Petroleum Institute (1981b).

Antimony: Inhalation can cause dermatitis, keratitis, conjunctivitis, and nasal septum ulceration. Amounts greater than about 0.1 g are considered to be lethal to humans by ingestion. Antimony shortens lifespan when fed to rats and mice. It also causes phenmonitis and heart and liver damage.

Arsenic: Chronic exposure to arsenic can lead to weakness, anorexia, bronchitis, gastrointestinal disturbances, peripheral neuropathy, skin disorders, and damage to the liver, heart, nerves, and kidneys. Exposure to arsenic compounds in drugs, food, and water have been causally associated with the development of cancer, primarily of the skin and lungs, although a direct connection has never been proven. Low doses stimulates plant growth. Arsenic may impact embryo and larval states of fish and benthic invertebrates.

Barium: Barium ion is a muscle poison causing stimulation and then paralysis. Soluble barium salts are skin and mucous membrane irritants. In animals, BaO and $BaCO_3$ cause paralysis. Ba is also poisonous to most plants. The barium ion is a physical antagonist of potassium, interfering with the vital cellular use of potassium.

Although elemental barium is extremely toxic, the barium compounds encountered during drilling and production activities are relatively nontoxic. The most commonly found form of barium is barium sulfate, which is insoluble in geochemical conditions and is not taken up by plants. Barium sulfate is not absorbed by animals or humans if ingested. It is commonly used internally for medical applications using X-ray diagnostics. Barium carbonate is moderately soluble and is more toxic, but is rarely used.

Cadmium: Excess exposure to cadmium can lead to renal failure, anemia, bone fractures, kidney stones, osteomalacia, retarded growth, pulmonary emphysema, and pain in the back and joints. Cadmium has been implicated in respiratory tract cancer. There is little evidence of carcinogenicity for exposure by ingestion. Organometallic derivatives may be concentrated in lipid tissues and cause chromosome damage. Cadmium interferes with the metabolism of zinc and copper in humans.

Chromium: The toxicity of chromium depends primarily on its chemical valence state and its concentration. Chromium is considered to be an essential element in humans at low levels. At higher concentrations, hexavalent chromium can be highly toxic, while trivalent chromium is relatively nontoxic. Hexavalent chromium can cause severe irritation to the respiratory system, asthma, and kidney damage. Some hexavalent chromium compounds are carcinogenic. Prolonged inhalation of trivalent chromium may cause scarring of the lungs. Other effects of chronic exposure at high levels include lung cancer, dermatitis, ulceration of the skin, chronic catarrh, and emphysema. However, virtually all chromium found in the drilling and production industry is in the low solubility, low toxicity trivalent form.

Bioassays on freshwater organisms for trivalent chromium at concentrations around 1.0 mg/l yielded a mixture of no effects to mixed sublethal effects after exposures of up to three months. A three-week LC_{50} value for *Daphnia magna* (freshwater shrimp) was reported as 2.0 mg/l. No significant mortalities were observed on *Neanthes arenaceodentata* (marine polychaetes) for three-week exposures to trivalent chromium at concentrations up to 12.5 mg/l. Bioassays on marine organisms for trivalent chromium yielded 96-hr LC_{50} values of 53 mg/l on juvenile fish and 24-hr LC_{100} values (100% mortality) of around 50 mg/l on invertebrates. For comparison, 96-hr LC_{50} values on the same invertebrate species for hexavalent chromium was about 3.0 mg/l (American Petroleum Institute, 1981b).

Cobalt: This metal is essential to blue-green algae and some bacteria, fungi, and green algae, but there is little evidence of its essentiality to higher plants. Normal human intake is 0.002 mg/day, with toxic levels at 500 mg/day. Cobalt metal dust is more toxic than salts in inhalation. Higher concentrations cause dermatitis, heart and gastrointestinal tract disorders, and liver and kidney damage.

Copper: Inhalation of dust causes lung and gastrointestinal disturbances. It affects erythrocytes and the liver and irritates skin and mucous membranes.

Lead: Prolonged exposure induces toxic responses in the hemotological, neurological, and renal systems, leading to brain damage, convulsions, behavioral disorders, and death. There is some evidence that some soluble lead salts are carcinogenic in some animals, but there is little evidence of their carcinogenicity in humans. Organometallic derivatives may be concentrated in lipid tissues and cause chromosome damage. Some plants show retarded growth at 10 ppm. Subtoxic effects have been observed in microflora at 0.1 ppm.

Manganese: Pathological effects on nerve cells and the liver have been reported.

Mercury: Chronic exposure to mercury causes weakness, fatigue, anorexia, and disturbances of gastrointestinal functions. Following high exposures, tremors and spasms of the fingers, eyelids, lips, and even the whole body can occur. In severe cases, delirium and hallucinations may occur. Mercury exposure can damage the nervous system, kidneys, and liver. There is no evidence of mercury being carcinogenic in humans. Organometallic derivatives may be concentrated in lipid tissues and cause chromosome damage. Subtoxic effects have been observed in microflora at 0.1 ppm. Detrimental effects have been observed in aquatic ecosystems at 0.005 ppm.

Nickel: Exposure to nickel can cause a sensitization of the skin and allergic reactions in the respiratory tract. It has been associated with nasal and lung cancer, but carcinogenicity from ingestion has not been proven. The carcinogenicity of nickel compounds appears to depend on the solubility of the compounds. Organometallic derivatives may be concentrated in lipid tissues and cause chromosome damage.

Vanadium: Exposure from inhalation affects the eyes and respiratory system. At high exposure levels, damage to the lungs, liver, kidneys, and heart have been observed. No evidence of carcinogenicity has been observed.

Zinc: Inhalation of zinc oxide causes an influenza-like illness. Moderate exposures have little adverse effects on the lungs. No evidence has been obtained suggesting that zinc compounds are carcinogenic. Zinc is an essential element in the human metabolism and is required in low concentrations. It is toxic to plants above 400 ppm and lethal to fish and other aquatic animals at 1.0 ppm.

3.5 PRODUCTION CHEMICALS

The various chemicals used during production have a widely varying potential for environmental impact, depending on the chemical and its concentration.

The environmental impact of acids varies somewhat with acid type. All acids can be corrosive to eyes and skin. Hydrofluoric acid can be lethal if sufficient quantities are absorbed through the skin, inhaled, or ingested. Effects from chronic exposure to hydrofluoric acid include fluorosis (fluoride poisoning) and kidney or liver damage. Chronic exposure to hydrochloric acid can cause irritation to mucous membranes, erosion of teeth, and aggravation of respiratory conditions such as asthma. Little aquatic toxicity data are available for acids.

Pesticides vary in toxicity. Prolonged or repeated exposure may cause various systemic effects, including damage to the nervous and muscular systems. Some pesticides are carcinogenic. Exposure to some pesticides can be fatal.

Glycol can be fatal if ingested in quantities of about 100 ml. Lower doses may be irritating to the mouth, throat, and stomach and can cause disorders of or damage to the digestive tract. Repeated exposure can cause kidney, brain, or liver damage. Blood chemistry and blood cells can also be affected.

Bioassays have been conducted for a variety of production chemicals using different freshwater and saltwater organisms. Table 3-10 summarizes the typical concentrations of some chemicals used for different types of applications. This information includes typical ranges of chemical concentrations when used, concentrations when discharged for disposal, and the LC_{50} values (Hudgins, 1992). From this table, it can be seen that the toxicities of production chemicals vary widely.

More detailed toxicity data are summarized for scale inhibitors in Table 3-11, for biocides in Tables 3-12a and 3-12b, reverse emulsion breakers in Table 3-13, emulsion breakers in Table 3-14, corrosion inhibitors in Table 3-15, paraffin inhibitors in Table 3-16, surfactants in Tables 3-17a through 3-17e, coagulants in Table 3-18, foam breakers in Table 3-19, and gas treatment chemicals in Table 3-20. Because of varying test protocols, a direct comparison of the toxicities of these chemicals may not be valid.

3.6   DRILLING FLUIDS

Two methods have been used to evaluate the environmental impact of drilling fluids. First are bioassays conducted using various organisms placed in different concentrations of drilling fluids. Second are direct measurements of environmental impact following disposal of drilling fluids, either in reserves pits or by offshore dumping.

3.6.1   Bioassays of Drilling Fluids

Bioassays using mysid shrimp (*Mysidopsis bahia*) are currently required for the offshore discharge of drilling fluids in the United

Table 3–10
Toxicity of Production Chemicals

Chemical Application	Typical Concentration During Usage (ppm)	Typical Concentration as Discharged (ppm)	LC$_{50}$ (ppm)
Scale inhibitor	3–10[1]	3–10	1,200–>12,000
			90%>3,000
	5,000[2]	50–500	
Biocides	10–50[1]	10–50	0.2–>1,000
			90%>5
	100–200[3]	100–200	
Reverse emulsion breakers	1–25[1]	0.5–12	0.2–15,000
			90%>5
Emulsion breakers	50[5]	0.4–4	4–40
			90%>5
Corrosion inhibitors	10–20[4]	5–15	0.2–5
			90%>1
	10–20[5]	2–5	2–1,000
			90%>5
	5,000[2]	25–100	
Paraffin inhibitors	50–300	0.5–3	1.5–44
			90%>3
Surfactant cleaners	—	—	0.5–429
			90%>5

[1]*Concentration during continuous operation.*
[2]*Maximum concentration in returns after batch job.*
[3]*Maximum concentration of slug.*
[4]*Water-soluble chemical.*
[5]*Oil-soluble chemical.*

Source: from Hudgins, 1992.
Copyright SPE, with permission.

States (Ayers et al., 1985). In this test, the drilling fluids are first mixed with seawater at a ratio of one part drilling mud to nine parts seawater. The pH of the solution is adjusted to that near seawater (7.8–9.0) by adding acetic acid. The mixture is stirred for five minutes and allowed to settle for one hour. A portion of the fluid is filtered through

(text continued on page 114)

Table 3-11
Acute Toxicity of Scale Inhibitors (96-hr LC_{50}, mg/L)

Generic Chemical Type	Fresh Water	Salt Water
Amine phosphate ester	>1,000	>4,309
Phosphonate	3,700->10,125	1,676->10,125

Source: after Hudgins, 1992.
Copyright SPE, with permission.

Table 3-12a
Acute Toxicity of Biocides (LC_{50}, mg/L)

Generic Type	Generic Chemical Type	Fresh Water	Salt Water
Aldehydes	Glutaraldehyde (25%)	16.9–43	2.1–1,100
	Glutaraldehyde (50%)	11.5–23.7	—
	Formaldehyde	18–64	23–1,000
Formaldehyde mixtures	With heterocyclic polyamine	41.4–73.3	2.9–1,000
	With alkyldimethyl benzyl quaternary	1.79–2.24	12–290
Quaternary	Ethoxy quaternary	0.35–1.32	174–1,000
	Dicocoamine	0.42–1.7	0.4–34
Amine salt	Cocodiamine acetate	0.22–1.6	0.719–965
	Cocodiamine fatty acids	0.73–0.92	0.22–670
	Others	0.09–1.62	24–922
Amine	Alkyl propylene diamine+ 2 ethylhexanol	0.75-0.78	0.49–49
Others	Metronidazole	>100	180
	2,2-dibromo-3-nitrilopropionamide	4.5–8.15	2.8->1,000
	Dithiocarbamates	1.29	1.38
	Isothiazalin	40.6	66.1–4,000
	2,4,5-trichlorophenate	0.86–1.26	—
	Toxaphene pesticide	0.036–0.042	

Test lengths are for either 48 or 96 hours.
Source: after Hudgins, 1992.
Copyright SPE, with permission.

Table 3-12b
Toxicity of Biocides (15-minute microtox, EC_{50}, mg/L)

Biocide	Pure	With Oxygen Scavenger
Ammonium bisulfate	250–500	—
Chlorinated aromatic	0.33	0.52
Fatty diamine	3.9	1.6
Formulated fatty diamine oxygen scavenger mixture	—	2.4
Fatty amine	3.7	1.1
Polymeric biguanide hydrochloride	1.1	1.5
Organobromide	2.5	33

Source: after Whale and Whitham, 1991.
Copyright SPE, with permission.

Table 3-13
Toxicity of Reverse Emulsion Breakers (96-hr LC_{50}, mg/L)

Generic Chemical Type	Fresh Water	Salt Water
Cationic polyelectrolyte + metal salts	1.2–4.4	56
Polyamine ester + zinc salt	—	235–1,020
Polyacrylate	16,713	13,467–15,621
Cationic polyelectrolyte	1.2–4.4	>1,000

Source: after Hudgins, 1992.
Copyright SPE, with permission.

Table 3-14
Toxicity of Emulsion Breakers (96-hr LC_{50}, mg/L)

Generic Chemical Type	Fresh Water	Salt Water
Oxyalkylated dopropylene glycol	40	—
Phenol formaldehydes	5.26–25.4	3.56–28
Alkyl aryl sulfonate	6.7–7.5	10

Source: after Hudgins, 1992.
Copyright SPE, with permission.

Table 3-15
Toxicity of Corrosion Inhibitors (96-hr LC$_{50}$, mg/L)

Generic Chemical Type	Fresh Water	Salt Water
Amide/imadizoline	0.26–75	2.12–261
Amide/imadizoline + quaternary	1.2–1.3	1–5
Quaternary	1.5–2.8	—
Ammonium salts	—	5.96–116
Amines	0.86	1.98–710
Sulfonate	—	220
Phenanthradine	6.1	—
Pyridine salt + quaternary	2.26	—
Alkyl morpholines	—	800–1,055
Ammonium bisulfite	75–423	77–788
Sodium sulfite	7,000	—

Source: after Hudgins, 1992.
Copyright SPE, with permission.

Table 3-16
Acute Toxicity of Paraffin Inhibitors (96-hr LC$_{50}$, mg/L)

Generic Chemical Type	Fresh Water	Salt Water
Vinyl polymer	42	2.7
Sulfonate salt	17–25.1	37.4
Alkyl polyether + aryl polyether	—	1.55
Other	17–44	13.3–37.4

Source: after Hudgins, 1992.
Copyright SPE, with permission.

Table 3-17a
Toxicity of Surfactant Cleaners (96-hr LC$_{50}$, mg/L)

Generic Chemical Type	Fresh Water	Salt Water
Oxyalkylate	3.5–192	5.6–429
Alkoxylated phenol	48-106	56–410
Cationic (quaternary)	0.5	40
Glycol ether	183	—

Source: after Hudgins, 1992.
Copyright SPE, with permission.

Table 3-17b
Toxicity of Nonionic Surfactants (96-hr LC$_{50}$, mg/L)

Surfactant Type	Test Species	Toxicity
Fatty alcohol ethoxylates (C_{12}-C_{15})	*Pimephales promelas*	<3
Fatty alcohol ethoxylates (C_{10}-C_{12})	*Pimephales promelas*	64
	Pleuronectes platessa	7.5
	Crangon crangon	22
	Algae	5
Fatty alcohol ethoxylates (C_{10})	*Pimephales promelas*	20
	Pholis gunnellus	90
	Gasterosteus aculeatus	90
	Crangon crangon	180
	Chaetogammarus marinus	49
Di-sec-butyl-phenol ethoxylate	*Pimephales promelas*	50
Polypropylene glycols (MW=4,000)	*Pimephales promelas*	>100
Polypropylene glycols (MW=400)	*Pimephales promelas*	>100
Ethoxylated alkyl alcohols + methanol	*Crangon crangon*	33–100
Ethoxylated alkyl alcohols + isopropanol	*Crangon crangon*	100–330
Unspecified surfactant	Fish	10–40

Source: after Maddin, 1991.
Copyright SPE, with permission.

Table 3-17c
Toxicity of Anionic Surfactants (96-hr LC$_{50}$, mg/L)

Surfactant Type	Test Species	Toxicity
Dodecylbenzenesulfonic acid		
with di-sec-butylphenol ethoxylate	*Gasterosteus aculeatus*	0.32
Dodecyclbenzenesulfonic acid	*Crangon crangon*	330–1,000
Sodium dodecyl-benzenesulfonate	Fish	<10
	Salmo gairdneri	4–6
	Cyprinus carpio	4–6
Sodium tetrapropyl-benzenesulfonate	*Salmo gairdneri*	1–2
	Cyprinus carpio	1–2
Sodium alkyl-(branched) benzenesulfonate	*Cyprinus carpio*	18
Sodium alkyl-(C10-C15) benzenesulfonate	*Salmo gairdneri*	1.9
Disodium decyldiphenyl-ether		
disulfonate	*Pimephales promelas*	4
Sodium polynaphthalene-sulfonate	*Pimephales promelas*	375
Ammonium decyl		
poly-ethoxyether sulfate	*Salmo gairdneri*	140
Sodium dodecyl-ethoxyether sulfate	Fish	<10
Sodium palmitate	Fish	10–12
Sodium oleate	Fish	10–12
Sodium stearate	Fish	10–12

Source: after Maddin, 1991.
Copyright SPE, with permission.

Table 3-17d
Toxicity of Cationic Surfactants (96-hr LC$_{50}$, mg/L)

Surfactant Type	Test Species	Toxicity
Alkyl (C$_8$-C$_{18}$) di-(2-hydroxyethyl)		
benzyl ammonium chloride	*Salmo gairdneri*	5.4
Octadecyldimethyl ammonium chloride	*Salmo gairdneri*	4
Unspecified biocides	*Crangon crangon*	0.2–>1,000
Perfluorooctylsulfon-amidopropyltrimethyl		
ammonium iodide	*Pimephales promelas*	30
Dodecyltrimethyl ammonium chloride		
with dodecyldiethanol-amine oxide	*Crangon crangon*	55

Source: after Maddin, 1991.
Copyright SPE, with permission.

Table 3-17e
Toxicity of Amphoteric Surfactants (96-hr LC$_{50}$, mg/L)

Surfactant Type	Test Species	Toxicity
Dodecylbetaine	*Pimephales promelas*	12
Dodecylbetanie with polypropylene glycol (MW=400)	*Pimephales promelas*	87
	Daphnia magna	11

Source: after Maddin, 1991.
Copyright SPE, with permission.

Table 3-18
Toxicity of Coagulants (96-hr LC$_{50}$, mg/L)

Generic Chemical Type	Fresh Water	Salt Water
Polyamine ester	—	>1,000
Polyacrylamide	—	14,800
Phosphate ester	—	1,800
Polyamine quaternary	0.24–0.52	—
Polyquaternary	498	21

Source: after Hudgins, 1992.
Copyright SPE, with permission.

Table 3-19
Acute Toxicity of Foam Breakers (96-hr LC$_{50}$, mg/L)

Generic Chemical Type	Fresh Water	Salt Water
Alcohol modified fatty acid	50	—
Trybutyl phosphate	—	8.5

Source: after Hudgins, 1992.
Copyright SPE, with permission.

Table 3-20
Toxicity of Gas Treatment Chemicals (96-hr LC$_{50}$, mg/L)

Chemical	Fresh Water	Salt Water
Methanol	8,000->10,000[1]	12,000-28,000
Ethylene glycol	>10,000	>20,000[1]
Diethylene glycol	>5,000->32,000[2]	—
Triethylene glycol	>10,000–62,600	>1,000[3]

[1]*48-hour test*
[2]*24-hour test*
[3]*23-day test*

Source: after Hudgins, 1992.
Copyright SPE, with permission.

(text continued from page 107)

a 0.45 micron filter and designated the "liquid phase." The remaining unfiltered fluid is designated the "suspended particulate phase." The settled material at the bottom of the mixing vessel is called the "solid phase." Chemical additives, if any, are then mixed with this liquid for the toxicity test.

Mysid shrimp are used as the test organisms for the liquid and suspended particulate phases, while hard-shell clams are used for the solid phase. The U.S. Environmental Protection Agency has set a mysid shrimp toxicity limit (96-hour LC$_{50}$) for drilling mud discharge into the United States outer continental shelf (OCS) waters of 30,000 ppm (3%) for suspended particulate phase (after the 9:1 dilution with seawater). Materials with LC$_{50}$ values greater than one million ppm do not kill at least one half of the mysid shrimp during the 96-hour test.

Testing drilling fluids for toxicity in aquatic systems is difficult, however, because much of the material settles rapidly, and what remains suspended may partition into two or three discrete layers. This makes repeatability of the exact test conditions difficult. The condition of the test animals prior to exposure to the drilling fluids is also an important factor in determining the test repeatability. Laboratory tests, however, have shown similar results from different labs for the toxicity of various materials in drilling fluids (Parrish and Duke, 1988).

To speed permitting of new offshore wells and eliminate the need for bioassays on every drilling fluid prior to discharge, a set of eight generic drilling muds were developed for offshore use in the United

States (Ayers et al., 1985; Arscott, 1989). These muds have been approved for use in specific regions without bioassay testing every time a mud is to be discharged. Special chemical additives like lost circulation materials and lubricants can also be used if they come from an approved additive list. The discharge of diesel or free oil is not permitted under the generic mud program, although cuttings contaminated with oil can be discharged if they are washed and do not cause a sheen. If generic muds are not used, permits must be obtained on a well-by-well basis under the National Pollutant Discharge Elimination System (NPDES).

A number of bioassay studies have been conducted to determine the toxicity of various drilling muds and their additives. Two sets of toxicity data for these generic muds are given in Tables 3.21a and 3.21b. From these tables, it can be seen that the generic muds generally pass the 30,000 ppm toxicity limit on the liquid phase. The 95% confidence limit on the measured LC_{50} toxicities from one set of mysid shrimp bioassays has been reported to be about 30% of the measured value (Parrish et al., 1989). Toxicity data from several nongeneric muds are given in Table 3-22. From this table, it can be seen that muds that

Table 3-21a
Toxicity of Generic Drilling Muds (96-hr LC_{50}, ppm)

Generic Mud Type	Liquid Phase Toxicity[1]	Suspended Particulate Phase Toxicity[1]
Potassium chloride polymer	58,000–66,000	25,000–70,900
Lignosulfonate seawater	283,500–880,000	53,200–870,000
Lime	393,000–>1,000,000	66,000–860,000
Nondispersed	>1,000,000	>1,000,000
Spud mud (slugged intermittently with seawater)	>1,000,000	>1,000,000
Seawater/freshwater gel	>1,000,000	>1,000,000
Lightly treated lignosulfonate freshwater/seawater	>1,000,000	>1,000,000
Lignosulfonate freshwater	>1,000,000	506,000–>1,000,000

[1]*Mysid shrimp*

Source: from Ayers et al., 1985.
Copyright SPE, with permission.

Table 3-21b
Toxicity of Generic Drilling Muds (96-hr LC$_{50}$, ppm)

Generic Mud Type	Liquid Phase Toxicity[1]
Potassium chloride polymer	27,000
Lignosulfonate seawater	516,000
Lime	163,000
Nondispersed	>1,000,000
Spud mud (slugged intermittently with seawater)	>1,000,000
Seawater/freshwater gel	>1,000,000
Lightly treated lignosulfonate freshwater/seawater	654,000
Lignosulfonate freshwater	293,000

[1]*Mysid shrimp*

Source: from Arscott, 1989.
Copyright SPE, with permission.

Table 3-22
Mysid Shrimp Toxicity of Drilling Mud Additives

Mud Type	96-hr LC$_{50}$ (ppm)
Potassium chloride polymer (generic #1)	33,000
Lignosulfonate seawater (generic #2)	621,000
Lime (generic #3)	203,000
Lignosulfonate freshwater (generic #8)	300,000
PHPA 9.6 lbm/gal	>1,000,000
PHPA 14.3 lbm/gal	>1,000,000
PHPA/20% NACL/14.5 lbm/gal	140,000
PHPA seawater 13.5 lbm/gal	>1,000,000
Cationic Mud	>1,000,000
Freshwater chrome-lignosulfonate+2% diesel	5,970
Freshwater chrome-lignosulfonate+2% mineral oil (15% aromatics)	4,740
Freshwater chrome-lignosulfonate+2% mineral oil (0% aromatics)	22,500
Mineral oil	180,000

Source: from Wojtanowicz, 1991.
Copyright SPE/IADC, with permission.

contain chrome lignosulfonate and oil may fail the 30,000 ppm require-ment for offshore discharge. Bioassay on many commercial drilling fluid additives have also been conducted (*Offshore*, 1991a and 1991b).

Oil-based muds using diesel are more toxic than those using mineral oils. Studies have shown that the toxicity of mineral oils can be 5 to 14 times lower than diesel (Wojtanowicz, 1991). The mechanisms of toxicity reduction has been attributed to a reduced content of aromatic hydrocarbons in mineral oils and a low water solubility of the toxic components that are present. Diesel oil typically has between 30% and 60% aromatic compounds, while some mineral oils have virtually no aromatic compounds.

Conklin and Rao (1984) reported that the toxicity of whole drilling fluid on grass shrimp varies significantly with its formulation. The addition of diesel oil to the drilling fluids at a level of 0.9% increased the toxicity to grass shrimp by a factor of about 200, while the addition of mineral oil at the same concentration increased toxicity by a factor of about 50.

One of the difficulties with conducting bioassays on drilling muds is that new additives and formulations are continually being developed. The high cost of bioassays makes it difficult to justify bioassays on all conceivable combinations of additives and formulations. One approach that has been suggested to minimize the number of bioassays conducted is to measure the toxicity of the individual additives and then use an appropriate mathematical model to estimate the toxicity of their combinations. One mathematical model that has been proposed is to add the mass weighted reciprocals of the LC_{50} values of all constituents.

$$\frac{1}{LC_{50_{mixture}}} = \sum_i \frac{x_i}{LC_{50_i}} \tag{3-5}$$

where x_i is the mass fraction of the **i** component. Toxicity measure-ments on additives and their combinations have shown that this model results in calculated LC_{50} values for mixtures that are significantly lower than those measured, i.e., the mixture is less toxic than predicted by this formula (Parrish et al., 1989).

Drilling fluids can have significant sublethal effects on marine organisms. Parrish and Duke (1990) have summarized the work of a number of studies in this area. Sublethal effects of ferrochrome lignosulfonate were observed on corals at levels of 0.1 ml/l of used drilling fluid in seawater. Lobsters were observed to have an inhibited response to food odors at drilling fluid concentrations as low as 0.01 mg/l, and lethality (96-hr LD_{50}) was observed for lobster larvae at concentrations between 0.074 and 0.5 ml/l for various drilling fluids. Behavioral changes, including delays in feeding, molting, and shelter construction, were observed at levels as low as 0.007 mg/l. Drilling fluid concentrations between 1 and 10 mg/l adversely effected fertilization and subsequent embryo development of estuarine minnows; but the concentration where an effect was observed varied significantly with the particular drilling fluid tested. Sea urchins showed reduced fertilization rates when exposed to 223 mg/l barium sulfate. Behavioral characteristics, such as foraging by fish, gaping by scallops, and burrowing by shrimp, however, were unaffected by what was considered a realistic deposition rate on the sea floor within a 50-meter radius of a drilling platform.

Some accumulation of barium and chromium from the solids portion of used lignosulfonate drilling fluids has been observed in some benthic (sea bottom dwelling) species following exposure (American Petroleum Institute, 1985d). Some sublethal impacts were observed that included alterations in biochemical composition, depletion of micronutrients, and altered respiration and excretion rates. Once contaminated animals were placed in a clean environment, however, the concentrations in the animals was reduced to nominal levels. In other species, however, there were no observed bioaccumulation or effects.

The concentrations of drilling fluids that had no observable effect on the development of embryos of estuarine minnows (*Fundulus heteroclitus*), sand dollars (*Echinarachnius parma*) and sea urchins (*Strongylocentrotus purpuratus, Lytechinus pictus,* and *L. variegatus*) were measured by the U.S. Environmental Protection Agency (1983). Fish embryos were placed in the liquid phase of drilling fluids one minute after fertilization and maintained for the duration of their development. Sand dollar and sea urchin embryos were placed in the test medium 10–15 minutes after fertilization and kept there for 96 hours. The "safe" concentration—the concentration that is 10% of the lowest concentration that had an observable effect—was measured and is reported in Table 3-23. These safe concentrations were typically 1–100 microliters per liter.

Table 3-23
Toxicity of Drilling Fluids to Hard Clams, EC$_{50}$

Drilling Fluid Type[1]	Concentration (microl/l)
Seawater lignosulfonate	100
Seawater lignosulfonate	1
Seawater lignosulfonate	10
Lightly treated lignosulfonate	10
Freshwater lignosulfonate	100
Lime	10
Freshwater lignosulfonate	100
Freshwater/ seawater lignosulfonate	100
Reference drilling fluid	1

[1]*Duplicate drilling fluid types are from different formulations.*
Source: U.S. Environmental Protection Agency, 1983.

3.6.2   Impact of Drilling Fluid Disposal

Drilling fluids used for onshore wells are primarily disposed of in reserves pits, while in many areas drilling fluids from offshore platforms have been dumped overboard. A number of studies have been conducted on the impact of these discharges.

For most drilling muds, sodium has the greatest potential to impact the environment from the onshore disposal in reserves pits (Miller, 1978). Heavy metals are also of concern, although their potential to leach away from the pit and contaminate the groundwater is limited by their low concentration and low solubility (Mosley, 1983; Branch et al., 1990; Crawley and Branch, 1990; Candler et al., 1990; American Petroleum Institute, 1983). Extensive field studies have suggested that the onshore disposal of drilling wastes in reserves pits poses no serious threat to human health or the environment (American Petroleum Institute, 1983). In some cases, crop yield was improved following the disposal of drilling wastes.

A number of field studies have been conducted to measure the impact of discharging drilling fluids on the benthic community around offshore platforms. These studies have revealed elevated levels of hydrocarbons and heavy metals in the sediments surrounding platforms. Most of these hydrocarbons and heavy metals are associated with cuttings, making it possible to estimate the deposition of these materials by modeling the deposition of the cuttings. Models for sediment deposition following discharge from offshore platforms are available (MacFarlane and Nguyen, 1991).

In one study, the heaviest accumulations of hydrocarbons and heavy metals were found to be within about 100 meters of the platforms, with lower accumulations at farther distances (American Petroleum Institute, 1989c). The impact of these accumulations on the benthic community was uncertain. Seasonal variations in the organic matter content from nearby river runoff was greater than the concentrations from the platform. Seasonal variations in the benthic community were also greater than those observed at varying distances from the platform.

The greatest impact of offshore discharge of drilling fluids is when oil-based muds are used. Elevated hydrocarbon levels in the sediments and impacts on the benthic community have been measured at distances of several kilometers from platforms (Bakke et al., 1990; Peresich et al., 1991). The hydrocarbon concentration in the sediments, however, decreased significantly over a period of several years following discharge. The distance away from a platform that elevated levels of hydrocarbons can be detected may also depend on whether the cuttings were washed prior to discharge (De Jong et al., 1991a). The threshold level of hydrocarbons in subsea sediments below which no effects were observed on the mortality of the heart urchin (*Echinocardium ordatum*) was determined to be on the order of 10–100 mg oil/kg sediment (De Jong et al., 1991b). Because of these effects, the discharge of oil-based muds and their associated cuttings is prohibited in many areas around the world.

3.7 PRODUCED WATER

The potential for environmental impact following the discharge of produced water arises primarily from its high salt content, its heavy metals content, its dissolved or suspended hydrocarbons, and its oxygen deficiency.

The acute toxicity of a selection of produced waters to mysid shrimp (96-hr LC_{50}) was found to range between 1.3% to 9.3% by volume of produced water in seawater (U.S. Environmental Protection Agency, 1989). Sublethal effects were observed for produced water concentrations as low as 0.5% after 19 days of exposure. The toxicity of the produced water could not be correlated with total volatile organic carbon, total organic carbon, oil and grease, or salinity.

Field studies around offshore platforms have shown that the impact of produced water discharge depends on the volume of water discharged and the water depth. Except for shallow waters, little effect on the benthic community has been observed at distances greater than about 100 meters from the platform (American Petroleum Institute, 1989d; Rabalais et al., 1990).

Onshore discharges of produced water may be allowed if the water has a "beneficial use" in agriculture and wildlife propagation, even if it is not suited for human use. In Wyoming, for example, acceptable water quality is determined if more than 50% of water fleas and fathead minnows can survive in the produced water for 48 and 96 hours, respectively (Mancini and Stilwell, 1992).

3.8 NUCLEAR RADIATION

Humans are constantly exposed to a background level of nuclear radiation, from both natural and man-made sources. At most petroleum drilling and production facilities, there is no incremental radiation exposure from associated activities. At a few areas, however, naturally occurring radioactive materials (NORM) can accumulate to levels where a significant incremental exposure above background is possible.

3.8.1 Radioactive Decay

Radioactive decay occurs when the nucleus of an atom is in an unstable energy state. It is the process used by the nucleus to reach a more stable energy state. The three major types of radioactive decay are alpha, beta, and gamma decay. Other types of decay, such as spontaneous fission and spontaneous neutron emission, are possible but occur very infrequently. Induced neutron emission and induced fission can also occur when the nucleus has absorbed another particle, such as an alpha particle or a neutron.

Alpha decay is the emission of a helium nucleus (doubly ionized helium atom) from the nucleus of an unstable atom. Beta decay is the transformation of a neutron in the nucleus into a proton and an electron. The proton remains in the nucleus and the electron is emitted. In some cases of beta decay, a proton is transformed into a neutron

122   *Environmental Control in Petroleum Engineering*

and a positron (antielectron). The neutron remains in the nucleus and the positron is emitted. Gamma decay is the lowering of the energy of a nucleus through the emission of a photon of electromagnetic radiation. In most cases, gamma decay is of most concern in the petroleum industry.

Radioactive decay is the spontaneous change of a nucleus of an atom. Because it is a random process, there is no way to predict when a particular nucleus will decay. The decay of large numbers of atoms can be modeled through a decay probability, however. When a large number of nuclei are considered, the number of radioactive decay events is proportional to the number of nuclei present,

$$-\frac{dN}{dt} = \lambda N(t) \tag{3-6}$$

where λ is a constant of proportionality that depends on the type of nucleus and is a measure of the probability of decay for the nucleus. N is the number of nuclei present. If multiple decay modes are possible for a given nucleus, λ is the sum of the decay probabilities of each decay mode.

This equation can be solved for the number of nuclei as a function of time:

$$N(t) = N(0)e^{-\lambda t} \tag{3-7}$$

The most common measure of the rate of decay of radioactive nuclei is the time for half of the nuclei to decay. This time is called the *half-life* and can be expressed as

$$\frac{N(t)}{N(0)} = \frac{1}{2} = e^{-\lambda T} \tag{3-8}$$

where T is the half-life. The decay probability can be expressed in terms of the half-life, yielding the following equation for the number of nuclei as a function of time:

$$N(t) = N(0)\exp\left(-\frac{\ln(2)t}{T}\right) \tag{3-9}$$

The decay rate of a group of radioactive nuclei can also be expressed in terms of the total number of decay events per second, which is called *activity*. Activity is the primary measure of the radioactivity of a material. The units of activity are the Becquerel (Bq), which is equal to 1 decay/sec. A more common unit of activity is the Curie, which is equal to 3.7×10^{10} decays/sec.

A related measure of activity is the *specific activity*. For the specific activity, the concentration of radioactive nuclei is typically normalized in terms of activity per unit mass (for solids), activity per unit volume (for fluids), or activity per unit area (for surfaces).

3.8.2   Health Physics

The study of the effects of nuclear radiation on human health is the science of health physics. The effects of radiation are measured in terms of exposure or dose. Exposure is defined as the electrical charge released from ionization per unit mass of air. Dose is defined as the energy from the radiation absorbed per unit mass of material. One of the most widely used measures of radiation dose is the radiation absorbed dose (RAD), where

1 RAD = 100 erg/gram

The unit of RAD is not particularly useful for measuring human exposure because it neglects the biological effects of radiation. Different types of radiation have different biological effects for the same energy deposition. To account for these different biological effects, the RAD is multiplied by an empirical quality factor. The resulting value is called the dose equivalent and its most common unit is the REM (*roentgen equivalent man*). The quality factor for gamma radiation is 1 (one). Virtually all environmental impacts of nuclear radiation from the petroleum industry are from gamma radiation.

The impact of radiation exposure also depends on the type of radiation and where the source is located. The dose from alpha particles from a source external to the body is zero, because alpha particles cannot penetrate the skin and reach living cells. Beta particles are able to penetrate the surface layers of the skin and can provide a dose to living skin tissue. Any other exposure from alpha or beta particles can come only from ingesting or inhaling the radionuclide that emits the particle. Gamma rays, on the other hand, can penetrate through a body. Thus, the dose from gamma rays is a whole body dose, and all organs can be exposed. The dose from neutrons is very complex and will not be discussed.

There are two major types of biological effects of radiation: those affecting cells as a whole and those affecting the reproductive capacity of the cells.

The major effect on cells as a whole is for the radiation to break chemical bonds within the cell and create free radicals. The most common free radicals are those created from the decomposition of water, hydrogen and hydroxyl:

$$H_2O \rightarrow H^+ + OH^-$$

Hydroxyls can combine to form hydrogen peroxide:

$$OH^- + OH^- \rightarrow H_2O_2$$

Hydrogen peroxide is highly reactive and can react with most other molecules in the cell, disrupting the cellular chemistry.

The most important genetic changes involve cellular reproduction. If radiation breaks or alters the DNA molecules within a cell, the ability of the cell to replicate itself is impaired. In most cases, any alteration in the DNA prevents it from reproducing. In some cases, however, the cell is able to reproduce, but the subsequent cells may be mutated. Similar mutations in cells also arise from the elevated temperatures of cooking, from drugs, and from exposure to chemicals.

Experience has shown that a one-time whole body dose of less than about 50 REM will not result in any noticeable or measurable acute effects. A dose an order of magnitude higher, e.g., 400–500 REM, is a lethal dose for 50% of those receiving it. Thus, the LD_{50} for humans is about 500 REM, with death usually occurring about two months after exposure. A dose above 1,000 REM is considered lethal for all exposed, i.e., LD_{100} is about 1,000 REM.

Long-term effects from a single large exposure or from a chronic low-level exposure may include loss of hair, eye cataracts, cancer, or leukemia. Unfortunately, those effects also arise from causes other than nuclear radiation so it is difficult to determine whether or not they come from radiation exposure. In many cases, no effects will be observed following exposure to radiation.

Natural and man-made sources of nuclear radiation provide an average exposure of about 750 mREM/year per person. The natural background exposure of nuclear radiation varies widely but averages about 500 mREM/year. This exposure comes from cosmic rays, naturally occurring radioactive elements in the ground, the air (radon and carbon-14), and from naturally-occurring elements in our bodies and the foods we eat. Exposure from man-made radiation sources averages about 250 mREM/year. Man-made sources include medical and dental X-rays, smoking, color television, and luminous wristwatches. Nuclear power plants contribute less than 1 mREM/year. Actual exposure levels for any particular individual vary significantly.

The risks from nuclear radiation can also be placed in perspective by comparing the estimated loss of life expectancy to that from other health risks. As seen in Table 3-24, there is a finite risk, but it is small compared to many other risks.

The International Commission on Radiological Protection has set recommended exposure limits for radiation. The maximum recommended cumulative exposure to radiation is 5 REM per year. This level

Table 3-24
Estimated Loss of Life Expectancy from Health Risks

Health Risk	Loss of Life Expectancy (days)
Smoking 20 cigarettes/day	2,370 (6.5 years)
Overweight by 20%	985 (2.7 years)
All accidents combined	435 (1.2 years)
Auto accidents	200
Alcohol consumption	130
Home accidents	95
Drowning	41
Safest jobs (teaching)	30
Natural background radiation	8
Medical x-rays (U.S. average)	6
Natural catastrophes	3.5
1 rem occupational radiation dose	1
1 rem/year for 30 years	30
5 rem/year for 30 years	150

Source: Von Flatern, 1993.
Copyright Petroleum Engineer International, *with permission.*

is one tenth the level that causes medically observable changes in cellular chemistry. The maximum permitted occupational exposure is one tenth of the maximum recommended exposure level (500 mREM/yr), while the maximum permitted exposure to the general public is one tenth of the occupational level (50 mREM/yr). These limits do not include exposure from natural radiation or medical X-rays.

Radiation exposure limits are governed by the *as low as reasonably achievable* (ALARA) concept. Under ALARA, all exposures are kept to a minimum, even if the exposures are well below the maximum recommended levels.

3.8.3 Naturally Occurring Radioactive Materials

In most cases, the level of NORM found at a site and the subsequent dose from exposure are too low to represent a serious hazard to employees. At a few sites, however, the potential exists for exposures that exceed the recommended levels after only a few hours. The largest risk of NORM exposure is probably ingestion or inhalation of NORM by workers handling and cleaning contaminated equipment. Care must be taken to prevent buildup of NORM-contaminated scale on the ground after cleaning out equipment. Because of its long half-life (1,622 years), Ra-226 contaminated pipe yards could pose a health threat to future development of the area, particularly in urban areas.

To determine the level of NORM at a site, radioactive assays are conducted. The concentration of NORM in equipment or scale is important in determining whether the material is considered radioactive or not and how it can be disposed. These assays are expensive ($50–$150 per sample) and can take up to 90 days before the results become available (Miller et al., 1990).

Because of the cost and time required to assay NORM levels to determine handling and disposal options, several attempts have been made to develop a relationship between the specific activity of NORM to the levels of radiation around the equipment as measured by a hand-held detector (Carroll et al., 1990; Miller et al., 1990; Smith, 1987). Additional work in the area is needed, however.

3.9 AIR POLLUTION

The primary impacts of air pollutant from production activities comes from chronic exposure. For materials, the impact includes soiling or chemical deterioration of surfaces. For plants, the impact includes damage to chlorophyl and a disruption of photosynthesis. Sulfur dioxide can also accumulate in soils, lowering the pH and modifying the soil nutrient balance. The impact of air pollutants on humans and animals includes irritation and damage to respiratory systems.

The impact of sulfur dioxide and hydrocarbons (ethene) has been observed on plants at concentrations as low as 0.03 ppm and 0.05 ppm, respectively. Sulfur dioxide concentrations on the order of 1 ppm can cause constriction of airways in the respiratory tracts of humans (Seinfeld, 1986).

3.10   ACOUSTIC IMPACTS

Some of the operations associated with drilling and production can generate high noise (acoustic) levels. The impact of these noises, however, is normally small. The most important sources are the seismic operations used during exploration. A number of studies have been conducted on ways to minimize the environmental impact of these operations (Ruiz Soza, 1991; Wren, 1991; Wright, 1991; Bertherin, 1991).

An extensive review of the acoustic impact of drilling and production on marine mammals was conducted by the American Petroleum Institute (1989a). This review concluded that acoustic impacts from offshore petroleum operations, including sounds from ships, aircraft, seismic exploration, drilling, dredging, and production, are limited primarily to short-term responses by mammals. For example, an airplane flyby can cause pinnipeds (seals and walruses) to jump into the water, abandoning their young. No long-term impacts on marine mammal populations have been observed, however. Explosives can injure mammals in water within a few hundred meters, but seismic air guns are not believed to be physically harmful unless the animals are very close to the guns.

The effects of air guns on fish with swim bladders, e.g., anchovies, was also studied (American Petroleum Institute, 1987b). The overall effects of seismic surveys using air guns appears very small. Noticeable effects on eggs and larvae would only result from large numbers of multiple exposures to full seismic arrays. The largest reduction in survival rate (35%) was for four-day-old larvae exposed 3–4 times to air guns passing overhead at a distance of 10 feet. Seismic pulses with air guns appear to have a lethal radius for fish of about 1–2 meters.

The effect of chronic noise from underwater drilling on the behavioral and physiological responses of belukha whales has also been studied (American Petroleum Institute, 1986b). It was found that belukha whales, like other toothed cetaceans, have a hearing range of greatest sensitivity different from the frequency range of most industrial sounds. The only response to the sound observed was a startle response at the start of each playback session in a pool containing four whales. Thus, little effect on whale health or behavior is expected from drilling activities.

3.11 EFFECTS OF OFFSHORE PLATFORMS

When properly managed, the actual environmental impact of offshore exploration and production activities is very low. In some cases, the presence of offshore platforms can be beneficial. The subsea structure (jacket) provides a substrate for marine flora to grow. This growth is particularly important in areas where few rocks are found on the bottom to provide such a substrate, e.g., in the Gulf of Mexico or other deltaic systems. This flora then attracts fauna of different types and sizes. Eventually large fish are attracted to the platform, yielding a much higher fish concentration than is found in the open ocean. This high fish concentration provides enhanced commercial and recreational fishing opportunities.

When an offshore field is abandoned, the platform must be removed. The least expensive and safest method for platform removal has been to use explosives to sever the piles and conductor pipes below the mudline. The use of underwater explosives for this purpose, however, can be lethal to aquatic life swimming nearby. Monitoring of the surrounding area (within 1,000 yards) is now required in some areas before the charges can be detonated. Any endangered species in the area, such as sea turtles, must be removed before detonation. Other methods to sever the platform from its anchorage that have been considered include acid cutting, embrittlement through liquid nitrogen freezing, solid fuel cutting torches, water blasting, and mechanical cutters. These methods, however, may result in greater safety hazards to the personnel implementing them.

3.12 RISK ASSESSMENT

Risk assessment provides a numerical estimate of the probability of potentially adverse health effects from human exposure to environmental hazards. It identifies what the potential hazards may be, their potential impact, how many humans could be impacted, and what the overall impact might be.

Risk assessment can be used to identify and rank the substances that have the greatest potential environmental impact. This helps companies identify and prioritize efforts to ensure environmentally safe operations. Risk assessment studies also document environmentally responsible actions and can be used as a scientifically defensible study if litigation occurs. Risk assessment studies are expensive, however, and may not be feasible for small operations. They are normally required only for new emission sources or modified stationary sources. The calculations are complex and based on various exposure pathways. Sullivan (1991) provides a discussion of risk assessment for crude oil contamination.

Risk assessment consists of four steps: hazard identification, dose-response assessment, exposure assessment, and risk characterization.

Hazard Identification determines the nature and amount of toxic pollutants that could potentially be emitted. It identifies the potential adverse health effects associated with those pollutants. Hazard identification includes a qualitative review of the available information of each substance to determine which substances should be included in a detailed assessment. It also determines the potential exposure pathways for the spread of the pollutant following a release e.g., groundwater or airborne transport, and the affected populations. Information for hazard identification can be obtained from relevant federal, state, and local regulations, risk assessment studies from similar facilities, Material Safety Data Sheets, and technical journals.

Dose-Response Assessment determines the relationship between the magnitude of an exposure to a substance and the occurrence of specific health effects. It involves determining the actual toxicity of each substance identified in the hazard identification. Dose-response assessment includes obtaining a description of the toxic properties of the substances, including acute (short term) effects, noncarcinogenic chronic (long-term) effects, and the carcinogenic potential for different dose levels. The result of this assessment is a probability estimate of the incidence of the adverse effect as a function of human exposure level to the substance.

The hazards of noncarcinogenic substances are evaluated relative to an allowable daily exposure level called the reference dose. The reference dose is the maximum daily dose of a substance to which a human may be exposed and not be adversely affected. In most cases, this dose is based on nontoxic exposure levels in animals that are extrapolated to humans with safety factors. This method assumes that exposures have a threshold below which no adverse effects will occur.

Carcinogenic substances are evaluated using a model for the probability of a human developing cancer. Either animal or human data (when available) are used in developing the probabilities. These substances are normally assumed to have no threshold levels, and all effects are extrapolated to zero exposure levels. Regulatory agencies establish quantitative limits for exposure that are based on the projected "excess cancer risk" caused by exposure to individual sources. Cancer risk is typically estimated for a lifetime exposure (70 years) and is expressed as a probability of developing cancer within a lifetime as in the term "one chance in one million." The U.S. Environmental Protection Agency has developed dose-response relationships for many compounds, but their values should be critically reviewed before being used because they change as new data become available.

Exposure Assessment determines the extent of potential human exposure to any emitted substances. Its goal is to accurately estimate both the dose that reaches the person (the administered dose) and the dose that reaches the target tissue within the body (the target dose). It quantifies all potential transport routes for each substance, e.g., groundwater or airborne transport, and considers three types of exposure—ingestion, inhalation, and dermal (skin) adsorption. Human exposures are reported as maximum daily doses for noncarcinogens and lifetime average daily doses for carcinogens.

Exposure assessment includes characterizing the emissions, modeling dispersion of the emissions, and quantifying the resulting exposures from each pathway. It estimates the probable magnitude, duration, timing, and route of exposure and the size and nature of the population exposed, and provides uncertainties for these estimates.

A critical part of exposure assessment includes working with the relevant regulatory agencies to ensure that the proper type of tests and measurements are conducted. Unfortunately, there are few standard exposure evaluation methods for most substances. The U.S. Environmental

Protection Agency has developed a set of air quality models that can be used for airborne pollutant transport.

Risk Characterization describes the nature, magnitude, and uncertainty of the health risks associated with each pollutant. It is the combination of the dose-response assessment and the exposure assessment. Risk characterization determines a quantitative estimate for the risk. This risk level can then be compared to a risk level that is considered to be insignificant. In humans, risk levels of one in ten thousand and one in one million are often used by regulatory agencies as benchmarks for acceptable risk levels.

The risk to the "maximum-exposed individual," i.e., the individual who receives the worst-case exposure scenario, and the more realistic risk to the general population should both be determined. Risk characterization should include a discussion of background levels of pollutants and risks associated with other activities, including the risks if nothing is done. Finally, risk characterization should be flexible and incorporate an honest evaluation of the uncertainties of the information used in the analysis.

Acceptable risk for carcinogens is normally determined in one of two ways. The most common approach is to calculate the maximum risk for an individual assuming an exposure level at the highest predicted long-term concentration. The goal of this approach is to limit excess lifetime cancer risks to a predetermined level. The second method is to estimate the aggregate incidence of potential excess cancer cases for the exposed population within the vicinity of the source. Risk assessment studies have uncertainties, particularly when conservative data are used. If more realistic data are used with Monte Carlo simulation, a more realistic estimate of risk can be obtained (Gordon and Cayias, 1993).

REFERENCES

American Petroleum Institute, "Fate and Effects of Polynuclear Aromatic Hydrocarbons in the Aquatic Environment," API Publication 4297, Washington, D.C., May 1978.

American Petroleum Institute, "Analysis of Mussel (*Mytilus californianus*) Communities in Areas Chronically Exposed to Natural Oil Seepage," API Publication 4319, Washington, D.C., May 1980.

American Petroleum Institute, "Fate and Effects of Experimental Oil Spills in an Eastern Coastal Marsh System," API Publication 4342, Washington, D.C., Sept. 1981a.

American Petroleum Institute, "The Sources, Chemistry, Fate, and Effects of Chromium in Aquatic Environments," Washington, D.C., Nov. 1981b.

American Petroleum Institute, "Summary and Analysis of API Onshore Drilling Mud and Produced Water Environmental Studies," API Bulletin D19, Washington, D.C., Nov. 1983.

American Petroleum Institute, "Effects of Petroleum Residues on Intertidal Organisms of Bermuda," API Publication 4355, Washington, D.C., 1984a.

American Petroleum Institute, "Fish and Offshore Oil Development," API Publication 875-59302, Washington, D.C., 1984b.

American Petroleum Institute, "Toxicity of Dispersed and Undispersed Prudhoe Bay Crude Oil Fractions to Shrimp, Fish, and Their Larvae," API Publication 4441, Washington, D.C., Aug. 1985a.

American Petroleum Institute, "Oil Effects on Spawning Behavior and Reproduction in Pacific Herring (*Clupea harengus pallasi*)," API Publication 4412, Washington, D.C., Oct. 1985b.

American Petroleum Institute, "Methods of Storage, Transportation, and Handling of Drilling Fluid Samples," API Publication 4399, Washington, D.C., March, 1985c.

American Petroleum Institute, "Chronic Effects of Drilling Fluids Discharged to the Marine Environment," API Publication 4397, Washington, D.C., June 1985d.

American Petroleum Institute, "Influence of Crude Oil and Dispersant on the Ability of Coho Salmon to Differentiate Home Water from Non-Home Water," API Publication 4446, Washington, D.C., Dec. 1986a.

American Petroleum Institute, "Underwater Drilling—Measurement of Sound Levels and Their Effects on Belukha Whales," API Publication 4438, Washington, D.C., March 1986b.

American Petroleum Institute, "Effects of Crude Oil and Chemically Dispersed Oil on Chemoreception and Homing in Pacific Salmon," API Publication 4445, Washington, D.C., June 1987a.

American Petroleum Institute, "Effects of Airgun Energy Releases on the Northern Anchovy," API Publication 4453, Washington, D.C., Dec. 1987b.

American Petroleum Institute, "Field Studies on the Reproductive Effects of Oil and Emulsion on Marine Birds," API Publication 4466, Washington, D.C., Oct. 1988.

American Petroleum Institute, "Effects of Offshore Petroleum Operations on Cold Water Marine Mammals: A Literature Review," API Publication 4485, Washington, D.C., Feb. 1989a.

American Petroleum Institute, "API Environmental Guidance Document: Onshore Solid Waste Management in Exploration and Production Operations," API, Washington, D.C., Jan. 1989b.

American Petroleum Institute, "Fate and Effects of Drilling Fluid and Cutting Discharges in Shallow, Nearshore Waters," API Publication 4480, Washington, D.C., Sept. 1989c.

American Petroleum Institute, "Fate and Effects of Produced Water Discharges in Nearshore Marine Waters," API Publication 4472, Washington, D.C., Jan. 1989d.

American Petroleum Institute, "Bioaccumulation of Polycyclic Aromatic Hydrocarbons and Metals in Estuarine Organisms," API Publication 4473, Washington, D.C., May 1989e.

American Petroleum Institute, "Rapid Bioassay Procedures for Drilling Fluids," API Publication 4481, Washington, D.C., March 1989f.

Arscott, R. L., "New Directions in Environmental Protection in Oil and Gas Operations," *J. Pet. Tech.,* April 1989, pp. 336–342.

Ayers, R. C., Jr., Sauer, T. C., Jr., and Anderson, P. W., "The Generic Mud Concept for NPDES Permitting of Offshore Drilling Discharges," *J. Pet. Tech.,* March 1985, pp. 475–478.

Baker, K. A., "The Effect of the Lodgepole Sour Gas Well Blowout on Coniferous Tree Growth: Damage and Recovery," paper SPE 23331 presented at the Society of Petroleum Engineers First International Conference on Health, Safety, and Environment, The Hague, Netherlands, Nov. 10–14, 1991.

Bakke, T., Gray, J. S., and Reiersen, L. O., "Monitoring in the Vicinity of Oil and Gas Platforms: Environmental Status in the Norwegian Sector in 1987-1989," Proceedings of the U.S. Environmental Protectional Agency's First International Symposium on Oil and Gas Exploration and Production Waste Management Practices, New Orleans, LA, Sept. 10–13, 1990, pp. 623–634.

Bertherin, G., "Seismic Techniques in Guatemala: An Approach Yielding a New Dimension to Environmental Protection," paper SPE 23520 presented at the Society of Petroleum Engineers First International Conference on Health, Safety, and Environment, The Hague, Netherlands, Nov. 10–14, 1991.

Bozzo, W., Chatelain, M., Salinas, J., and Wiatt, W., "Brine Impacts to a Texas Salt Marsh and Subsequent Recovery," Proceedings of the U.S. Environmental Protection Agency's First International Symposium on Oil and Gas Exploration and Production Waste Management Practices, New Orleans, LA, Sept. 10–13, 1990, pp. 129–140.

Branch, R. T., Artiola, J., and Crawley, W. W., "Determination of Soil Conditions that Adversely Affect the Solubility of Barium in Nonhazardous

Oilfield Waste," Proceedings of the U.S. Environmental Protection Agency's First International Symposium on Oil and Gas Exploration and Production Waste Management Practices, New Orleans, LA, Sept. 10–13, 1990, pp. 217–226.

Burnham, K. and Bey, M., "Effects of Crude Oil and Ultraviolet Radiation on Immunity Within Mouse Skin," *J. Toxicology and Environmental Health,* Vol. 34, 1991, pp. 83–93.

Burnham, K. and Rahman, M., "Effects of Petrochemicals and Ultraviolet Radiation on Epidermal IA Expression In Vitro," *J. Toxicology and Environmental Health,* Vol. 35, 1992, pp. 175–185.

Calabrese, E. J. and Kenyon, E. M., *Air Toxics and Risk Assessment.* Chelsea, Michigan: Lewis Publishers, Inc., 1991.

Candler, J., Leuterman, A., Wong, S., and Stephens, M., "Sources of Mercury and Cadmium in Offshore Drilling Discharges," paper SPE 20462 presented at the Society of Petroleum Engineers 65th Annual Technical Conference and Exhibition, New Orleans, LA, Sept. 23–25, 1990.

Carroll, J. F., Gunn, R. A., and O'Brien, M. S., "Naturally Occurring Radioactive Material Logging," paper SPE 20616 presented at the Society of Petroleum Engineers 65th Annual Technical Conference and Exhibition, New Orleans, LA, Sept. 23–25, 1990.

Conklin, P. J. and Rao, K. R., "Comparative Toxicity of Offshore and Oil-Added Drilling Muds to Larvae of *Palaemonetes intermedius,*" *Archives of Environmental Contamination and Toxicology,* Vol. 13, 1984, pp. 685–690.

Crawley, W. W. and Branch, R. T., "Characterization of Treatment Zone Soil Conditions at a Commercial Nonhazardous Oilfield Waste Land Treatment Unit," Proceedings of the U.S. Environmental Protection Agency's First International Symposium on Oil and Gas Exploration and Production Waste Management Practices, New Orleans, LA, Sept. 10–13, 1990, pp. 147–158.

De Jong, S. A., Zevenboom, W., Van Het Groenewoud, H., and Daan, R., "Short- and Long-Term Effects of Discharged OBM Cuttings, With and Without Previous Washing, Tested in Field and Laboratory Studies on the Dutch Continental Shelf, 1985–1990," paper SPE 23353 presented at the Society of Petroleum Engineers First International Conference on Health, Safety, and Environment," The Hague, Netherlands, Nov. 10–14, 1991a.

De Jong, S. A., Marquenie, J. M., Van't Zet, J., and Zevenboom, W., "Preliminary Results on Dose-Effect Relationships of Thermally Treated Oil-Containing Drilled Cuttings in Boxcosms," paper SPE 23355 presented at the Society of Petroleum Engineers First International Conference on Health, Safety, and Environment," The Hague, Netherlands, Nov. 10–14, 1991b.

Derkics, D. L. and Souders, S. H., "Pollution Prevention and Waste Minimization Opportunities for Exploration and Production Operations," paper SPE 25934 presented at the Society of Petroleum Engineers/Environmental

Protection Agency's Exploration and Production Environmental Conference, San Antonio, TX, March 7–10, 1993.

Deuel, L. E., "Evaluation of Limiting Constituents Suggested for Land Disposal of Exploration and Production Wastes," Proceedings of the U.S. Environmental Protection Agency's First International Symposium on Oil and Gas Exploration and Production Waste Management Practices, Sept. 10–13, New Orleans, LA, 1990, pp. 411–430.

Gordon, R. D. and Cayias, J. L., "An Approach to Resolve Uncertainty in Quantitative Risk Assessment," paper SPE 25959 presented at the Society of Petroleum Engineers/Environmental Protection Agency's Exploration and Production Environmental Conference, San Antonio, TX, March 7–10, 1993.

Hastings, L., Cooper, G. P., and Burg, W., "Human Sensory Response to Selected Petroleum Hydrocarbons," in *Applied Toxicology of Petroleum Hydrocarbons,* H. N. McFarland et al. (editors). Princeton: Princeton Scientific Publishers, 1984, pp. 255–270

Hoskin, S. J. and Strohl, A. W., "On-Site Monitoring of Drilling Fluid Toxicity," paper SPE 26005 presented at the Society of Petroleum Engineers/Environmental Protection Agency's Exploration and Production Environmental Conference, San Antonio, TX, March 7–10, 1993.

Hudgins, C. M., Jr., "Chemical Treatments and Usage in Offshore Oil and Gas Production Systems," *J. Pet. Tech.,* May 1992, pp. 604–611.

Jones, F. V., Rushing, J. H., and Churan, M. A., "The Chronic Toxicity of Mineral Oil—Wet and Synthetic Liquid—Wet Cuttings on an Estuarine Fish, *Fundulus grandis,*" paper SPE 23497 presented at the Society of Petroleum Engineers First International Conference on Health, Safety, and Environment, The Hague, Netherlands, Nov. 10–14, 1991.

Kaszuba, J. P. and Buys, M W., "Reclamation Procedures for Produced Water Spills from Coalbed Methane Wells, San Juan Basin, Colorado and New Mexico," paper SPE 25970 presented at the Society of Petroleum Engineers/Environmental Protection Agency's Exploration and Production Environmental Conference, San Antonio, TX, March 7–10, 1993.

Logan, T. H. and Traina, S. J., "Trace Metals in Agricultural Soils," in *Metals in Groundwater,* H. E. Allen, E. M. Perdue, and D. S. Brown, (editors). Chelsea, Michigan: Lewis Publishers, Inc., 1993, an imprint of CRC Press, Boca Raton, FL, pp. 311–312.

MacFarlane, K. and Nguyen, V. T., "The Deposition of Drill Cuttings on the Seabed," paper SPE 23372 presented at the Society of Petroleum Engineers First International Conference on Health, Safety, and Environment, The Hague, Netherlands, Nov. 10–14, 1991.

Macyk, T. M., Nikiforuk, F. I., and Weiss, D. K., "Drilling Waste Landspreading Field Trial in the Cold Lake Heavy Oil Region, Alberta, Canada,"

Proceedings of the U.S. Environmental Protection Agency's First International Symposium on Oil and Gas Exploration and Production Waste Management Practices, New Orleans, LA, Sept. 10–13, 1990, pp. 267–280.

Maddin, C. M., "Marine Toxicity and Persistence of Surfactants Used in the Petroleum Producing Industry," paper SPE 23354 presented at the Society of Petroleum Engineers First International Conference on Health, Safety, and Environment, The Hague, Netherlands, Nov. 10–14, 1991.

Mancini, E. R. and Stilwell, C. T., "Biotoxicity Characterization of a Produced Water Discharge in Wyoming," *J. Pet. Tech.,* June 1992, pp. 744–748.

Miller, R. W., "Effects of Drilling Fluids Components and Mixtures on Plants and Soils," API Project Summary: 1974–1977, 1978, p. 33.

Miller, H. T., Bruce, E. D., and Scott, L. M., "A Rapid Method for the Determination of the Radium Content of Petroleum Production Wastes," Proceedings of the U.S. Environmental Protection Agency's First International Symposium on Oil and Gas Exploration and Production Waste Management Practices, Sept. 10–13, New Orleans, LA, 1990, pp. 809–820.

Mosley, H. R., "Summary of API Onshore Drilling Mud and Produced Water Environmental Studies," paper SPE 11398 presented at the Society of Petroleum Engineers 1983 IADC/SPE Drilling Conference, New Orleans, LA, Feb. 20–23.

Mount, D. R., Gulley, D. D., and Evans, J. M., "Salinity/Toxicity Relationships to Predict the Acute Toxicity of Produced Waters to Freshwater Organisms," paper SPE 26007 presented at the Society of Petroleum Engineers/Environmental Protection Agency's Exploration and Production Environmental Conference, San Antonio, TX, March 7–10, 1993.

National Research Council, *Oil in the Sea: Inputs, Fates, and Effects,* Washington, D.C.: National Academy Press, 1985.

Offshore, "Drilling Fluid Product Directory: Part I," Sept. 1991a, p. 43.

Offshore, "Drilling Fluid Product Directory: Part II," Oct. 1991b, p. 62.

Parrish, P. R. and Duke, T. W., "Variability of the Acute Toxicity of Drilling Fluids to Mysids (*Mysidopsis bahia*)," American Society for Testing and Materials, Special Technical Publication 976, 1988.

Parrish, P. R. and Duke, T. W., "Effects of Drilling Fluids on Marine Organisms," in *Ocean Processes in Marine Pollution, Vol. 6, Physical and Chemical Processes: Transport and Transportation,* D. J. Baumgartner and I. W. Duedall (editors). Malabar, Florida: Krieger Publishing Co., 1990.

Parrish, P. R., Macauley, J. M., and Montgomery, R. M., "Acute Toxicity of Two Generic Drilling Fluids and Six Additives, Alone and Combined, to Mysids (*Mysidopsis bahia*)," in *Drilling Wastes,* F. R. Engelhard, J. P. Ray, and A. H. Gillam (editors). New York: Elsevier, 1989, pp. 415–426.

Peresich, R. L., Burrell, B. R., and Prentice, G. M. "Development and Field Trial of a Biodegradable Invert Emulsion Fluid," paper SPE/IADC 21935 presented at the 1991 Drilling Conference, Amsterdam, The Netherlands, March 11–14, 1991.

Proctor, N. H., Hughes, J. P., and Fischman, M. L., *Chemical Hazards of the Workplace,* New York: Van Nostrand Reinhold, 1989.

Rabalais, N. N., Means, J., and Boesch, D., "Fate and Effects of Produced Water Discharges in Coastal Environments," Proceedings of the U.S. Environmental Protection Agency's First International Symposium on Oil and Gas Exploration and Production Waste Management Practices, New Orleans, LA, Sept. 10–13, 1990, pp. 503–514.

Ruiz Soza, O., "Maturin East Seismic Program: Environmental Impact Assessment," paper SPE 23388 presented at the Society of Petroleum Engineers First International Conference on Health, Safety, and Environment, The Hague, Netherlands, Nov. 10–14, 1991.

Rushing, J. H., Churan, M. A., and Jones, F. V., "Bioaccumulation From Mineral Oil—Wet and Synthetic Liquid—Wet Cuttings in an Estuarine Fish, *Fundulus grandis,*" paper SPE 23350 presented at the Society of Petroleum Engineers First International Conference on Health, Safety, and Environment, The Hague, Netherlands, Nov. 10–14, 1991.

Ryer-Power, J. E., Custance, S. R., and Sullivan, M. J., "Determination of Reference Doses for Mineral Spirits, Crude Oil, Diesel Fuel No. 2, and Lubricating Oil," paper SPE 26398 presented at the Society of Petroleum Engineers 68th Annual Technical Conference and Exhibition, Houston, TX, Oct. 3–6, 1993.

Seinfeld, J. H., *Atmospheric Chemistry and Physics of Air Pollution,* New York: John Wiley and Sons, 1986.

Smith, A. L., "Radioactive-Scale Formation," *J. Pet. Tech.,* June 1987, pp. 697–706.

St. Pe, K. M., Means, J., Milan, C., Schlenker, M., and Courtney, S., "An Assessment of Produced Water Impacts to Low-Energy, Brackish Water Systems in Southeast Louisiana: A Project Summary," Proceedings of the U.S. Environmental Protection Agency's First International Symposium on Oil and Gas Exploration and Production Waste Management Practices, New Orleans, LA, Sept. 10–13, 1990, pp. 31–42.

Sullivan, M. J., "Evaluation of Environmental and Human Risk from Crude-Oil Contamination," *J. Pet. Tech.,* Jan. 1991, pp. 14–16.

Tchobanoglous, G. and Burton, F. L., *Wastewater Engineering: Treatment, Disposal, and Reuse.* New York: McGraw Hill., Inc, 1991.

U.S. Bureau of Land Management, "Ecological Investigations of Petroleum Production Platforms in the Central Gulf of Mexico," BLM-YM-YM-P/T-81-018-3331, NTIS No. PB82-167834, 1981.

U.S. Environmental Protection Agency, "Effects of Drilling Fluids on Embryo Development," EPA 600/3-83-021, Washington, D.C., 1983.

U.S. Environmental Protection Agency, "Acute and Chronic Toxicity of Produced Water to Mysids (*Mysidopsis bahia*)," EPA/600/X-89/175, Washington, D.C., April 1989.

U.S. Salinity Staff, "Diagnosis and Improvement of Saline and Alkali Soils," *Agriculture Handbook* 68, U.S. Department of Agriculture, 1954.

Valkovic, V., *Trace Elements in Petroleum,* Tulsa: Petroleum Publishing Company, 1978.

Vickers, D. T., "Disposal Practices for Waste Waters from Coalbed Methane Extraction in the Black Warrior Basin, Alabama," Proceedings of the U.S. Environmental Protection Agency's First International Symposium on Oil and Gas Exploration and Production Waste Management Practices, Sept. 10–13, New Orleans, LA, 1990, pp. 255–266.

Von Flatern, R., "NORM Contamination Regulations Threaten Industry Economy," *Petroleum Engineer International,* May 1993, pp. 36–39.

Whale, G. F. and Whitham, T. S., "Methods for Assessing Pipeline Corrosion Prevention Chemicals on the Basis of Antimicrobial Performance and Acute Toxicity to Marine Organisms," paper SPE 23357 presented at the Society of Petroleum Engineers First International Conference on Health, Safety, and Environment, The Hague, Netherlands, Nov. 10–14, 1991.

Wojtanowicz, A. K., "Environmental Control Potential of Drilling Engineering: An Overview of Existing Technologies," paper SPE/IADC 21954 presented at the Society of Petroleum Engineers 1991 Drilling Conference, Amsterdam, The Netherlands, March 11–14, 1991.

Wojtanowicz, A. K., Shane, B. S., Greenlaw, P. N., and Stiffey, A. V., "Cumulative Bioluminescence—A Potential Rapid Test of Drilling Fluid Toxicity: Development Study," *SPE Drilling Engineering,* March 1992, pp. 39–46.

Wren, J. M., "Minimizing the Environmental Impact of Seismic Operations in Canada and Alaska," paper SPE 23386 presented at the Society of Petroleum Engineers First International Conference on Health, Safety, and Environment, The Hague, Netherlands, Nov. 10–14, 1991.

Wright, N. H., "Optimal Environmental Strategies: Fit for Exploration," paper SPE 23387 presented at the Society of Petroleum Engineers First International Conference on Health, Safety, and Environment, The Hague, Netherlands, Nov. 10–14, 1991.

CHAPTER 4

Environmental Transport of Petroleum Wastes

The environmental impact of most releases of petroleum industry wastes would be minimal if the wastes remained at their points of release. Unfortunately, wastes can migrate away from a release point by a number of pathways. These pathways include transport along the surface of the earth or along the surface of a body of water, transport through the soil through the pore structure, and transport through the air. These migration pathways are briefly discussed below.

4.1 SURFACE PATHS

Surface pathways of transport are those where the released material travels along either the soil or open water surface. Surface transport of petroleum wastes from releases on land occurs primarily when high volumes of liquid wastes are discharged onto the ground or when stormwater sweeps through a site. These liquids then flow down topographical drainage features until they either mix with existing surface waters, evaporate, or enter the pore network of the earth they flow over. Dikes and diversion trenches can be used to control such surface migration.

Surface transport of petroleum wastes on open water can occur with hydrocarbons because they are lighter than water. This transport of hydrocarbons will be controlled by natural water currents and wind. Because virtually all natural water currents are parallel to the shoreline, the primary direction of transport will be parallel to the shoreline. If an onshore wind blows across the hydrocarbons, they can be pushed to shorelines. Hydrocarbons spilled on water will either evaporate, enter the water column, ground on the shore, or be naturally degraded.

4.2   SUBSURFACE PATHS

Subsurface pathways of transport are those where released liquids enter the pore structure of soil or sinks below the surface of open waters.

4.2.1   Releases on Land

There are two primary types of subsurface transport for onshore releases that can impact the environment: the transport of fluids at or above the water table from surface spills and the transport of fluids from one geologic formation to another through improperly plugged and abandoned wells.

When petroleum industry materials are discharged onto the ground, the liquid fraction, including any dissolved chemicals, begins to enter the pore network. These materials can travel through soil pore network in four ways. First, a separate nonaqueous phase liquid (NAPL) can flow through the pores. Second, contaminants can dissolve into groundwater and be transported by it. Third, very small solids (colloids) can also be transported with the water, although large particles will be filtered by the porous media. Fourth, volatile contaminants can be transported as a vapor through the vadose (air saturated) zone.

The transport of wastes through groundwater depends on a number of factors, including the permeability of the soil, capillary pressure between phases in the soil, solubility of the waste, partitioning coefficients, adsorption properties, and volatility. Adsorption, partitioning and volatilization decrease the concentration of chemicals in water, while leaching, desorption, and runoff increase the concentration. A review of the mechanisms of hydrocarbon transport in groundwater has been presented by Hunt et al. (1988a, 1988b).

Metals tend to form insoluble complexes in high-pH environments, minimizing their ability to leach away from a site (American Petroleum Institute, 1983b). The primary mechanisms for the fixation of metals by soils are absorption, ion exchange, and chemical precipitation. Ion exchange and adsorption are surface phenomena that are highly dependent on soil type and composition, particularly the amount of clays present. Factors that affect adsorption are the structural characteristics of the chemical, the organic content of the soil, the pH of the fluid medium, the soil grain size, the ion exchange capacity of the soil (clay content), and the temperature. Migration of heavy metals away from drill sites generally does not occur.

A number of numerical models having different levels of capabilities are available (American Petroleum Institute, 1986, and American Petroleum Institute, 1988). Unfortunately, most models neglect capillary trapping of the oil and air and hysteresis of relative permeability. Monte Carlo models allowing multiple realizations of possible contaminant transport have also been developed (Parker et al., 1993).

Another important pathway for the transport of petroleum wastes is improperly plugged and abandoned wells. These wells allow fluids from geologic formations having high hydrocarbon, salt, and/or heavy metals concentrations to flow into formations containing fresh water. Wells that are properly plugged and abandoned do not provide a permeable flow channel for fluids. Fluid flow, however, is not possible between layers if they are in hydrostatic pressure equilibrium, regardless of whether channels exist between the layers.

Numerical modeling of fluid flow in improperly abandoned wells can indicate the likelihood of freshwater contamination at a particular site (Warner and McConnell, 1993). The relative contamination potential of abandoned wells ranges from highly likely to impossible, depending on the age of the well, the depth of the well, the type of well, how the well was constructed, how it was plugged, the history of well activity, and the hydrogeologic conditions at the site.

4.2.2 Releases on Water

Transport of petroleum wastes below the surface of water depends primarily on the currents in the water and the topography of the floor of the water body. Produced waters typically have a greater salinity than fresh water or seawater, making them more dense. Discharged produced waters then sink until they either reach a density equilibrium with the seawater or reach the sea floor. Numerical models have been developed to model the transport of discharged drilling muds and produced water (Arscott, 1989). Two such models are the EPA's CORMIX1 and the Offshore Operators Committee models.

4.3   ATMOSPHERIC PATHS

Many petroleum industry wastes are gaseous and will be dispersed into the air, where they are transported with the wind. Upon release, airborn pollutants undergo transport by wind (advection), dispersion from atmospheric turbulence, and removal from deposition on the ground, vegetation, and buildings. Chemical transformations may also take place that alter the chemical and/or physical state of the emitted pollutant. Onshore sources of air pollutants are generally regulated by the total emission rates, while offshore sources are generally regulated so that the resulting onshore levels of pollutants are below specified levels.

To obtain permits to emit air pollutants for many applications, air-quality modeling is required (Sheehan, 1991). Such modeling relates the downwind concentration of released pollutants to their emission rates. Computer-based models are available that use information on the emission rate, physical characteristics of the emission source, the topography of the terrain over which the pollutants travel, and the meteorological conditions of the area to calculate the pollutant concentration downwind of the source (Moroz, 1987; Smith, 1987; American Petroleum Institute, 1983a; American Petroleum Institute, 1984; American Petroleum Institute, 1985a; American Petroleum Institute, 1985b). A discussion of models accepted by the U.S. Environmental Protection Agency is available (U.S. Environmental Protection Agency, 1986).

REFERENCES

American Petroleum Institute, "Model Performance Evaluation for Offshore Releases," API Publication 4387, Washington, D.C., Dec. 1983a.

American Petroleum Institute, "Summary and Analysis of API Onshore Drilling Mud and Produced Water Environmental Studies," API Bulletin D19, Washington, D.C., Nov. 1983b.

American Petroleum Institute, "Dispersion of Emissions from Offshore Oil Platforms—A Wind-Tunnel Modeling Evaluation," API Publication 4402, Washington, D.C., May 1984.

American Petroleum Institute, "Plume Rise Assessment Downwind of Oil Platforms for Neutral Stratification," API Publication 4420, Washington, D.C., Dec. 1985a.

American Petroleum Institute, "Development and Application of a Simple Method for Evaluating Air Quality Models," API Publication 4409, Washington, D.C., Jan. 1985b.

American Petroleum Institute, "Review of Ground-Water Models," API Publication 4434, Washington, D.C., 1986.

American Petroleum Institute, "Phase Separated Hydrocarbon Contaminant Modeling for Corrective Action," API Publication 4474, Washington, D.C., Oct. 1988.

Arscott, R. L., "New Directions in Environmental Protection in Oil and Gas Operations," *J. Pet. Tech.,* April 1989, pp. 336–342.

Hunt, J. R., Sitar, N., and Udell, K. S., "Nonaqueous Phase Liquid Transport and Cleanup: 1. Analysis of Mechanisms," *Water Resources Research,* Vol. 24, No. 8, Aug. 1988a, pp. 1247–1258.

Hunt, J. R., Sitar, N., and Udell, K. S., "Nonaqueous Phase Liquid Transport and Cleanup: 2. Experimental Studies," *Water Resources Research,* Vol. 24, No. 8, Aug. 1988b, pp. 1259–1269.

Moroz, W. J., "Air Pollution Concentration Prediction Models," in *Air Pollution,* E. E. Pickett. New York: Hemisphere Publishing Company, 1987.

Parker, J. C., Kahraman U., and Kemblowski, M. W., "A Monte Carlo Model to Assess Effects of Land-Disposed E&P Waste on Groundwater," paper SPE 26383 presented at the Society of Petroleum Engineers 68th Annual Technical Conference and Exhibition, Houston, TX, Oct. 3–6, 1993.

Sheehan, P. E., "Air Quality Permitting of Onshore Oil and Gas Production Facilities in Santa Barbara County, California," paper SPE 21767 presented at the Society of Petroleum Engineers Western Regional Meeting, Long Beach, CA, March 20–22, 1991.

Smith, B. P., "Exposure and Risk Assessment," in *Hazardous Waste Management Engineering,* E. J. Martin, and J. H. Johnson, Jr. (editors). New York:Van Nostrand Reinhold Company, Inc., 1987.

U.S. Environmental Protection Agency, "Guidelines on Air Quality Models (Revised)," EPA-450/2-78R, Research Triangle Park, NC, 1986.

Warner, D. L. and McConnell, C. L., "Assessment of Environmental Implications of Abandoned Oil and Gas Wells," *J. Pet. Tech.,* Sept. 1993, pp. 874–880.

CHAPTER 5

Planning for Environmental Protection

Many operations in the petroleum exploration and production industry have the potential to impact the environment in some way. Because of the high costs of noncompliance with the numerous regulations governing the industry and the high costs associated with the loss of public trust for damaging the environment, substantial resources must be dedicated to minimizing environmental impact. Because industry resources are limited, comprehensive environmental protection plans, including waste management and contingency plans, are needed to optimize the use of those resources.

One of the first steps in developing environmental protection plans is to conduct an environmental audit to identify all of the waste streams at a particular site and to determine whether those waste streams are being handled in compliance with all applicable regulations. Once an audit has been conducted, a written waste management plan for managing each waste stream should be developed. These plans identify how each waste stream is to be handled, stored, transported, treated, and disposed. The plan should also indicate how records are to be kept. Contingency plans are needed to minimize the impacts of accidental releases of materials and should incorporate relevant emergency responses. Several benefits of environmental audits and waste management plans are that they:

1. Ensure compliance with applicable environmental laws and regulations at a reasonable cost.
2. Minimize environmental damage from operations.

3. Minimize short- and long-term liabilities and risks associated with facilities operations.
4. Minimize operating costs through savings in raw materials and production costs.
5. Minimize personnel costs associated with waste management by having a written plan available.
6. Minimize costs of treating and disposing of wastes.
7. Minimize employee exposure to potentially hazardous materials.
8. Maintain a favorable corporate image.

Environmental protection plans should be developed with the guidance of people who are knowledgeable in the technical, regulatory, and operational aspects of systems operations and waste disposal. To be successful, these plans need the visible support of top management and require the active participation of field personnel, both in developing and implementing them. Because operations, regulations, and technology are constantly changing, environmental audits should be conducted periodically and associated waste management and contingency plans should be updated as needed.

An assessment of the potential environmental impact from future developments should also be conducted, and may be required in some areas. Such assessments include identifying all areas that the development may impact, quantifying the scale of that impact, and comparing it to regulatory standards. The findings of this assessment can be used to improve the design of facilities to reduce associated environmental risks. The entire project should be reevaluated at regular intervals to ensure minimal environmental impact (Grogan, 1991).

5.1 ENVIRONMENTAL AUDITS

An important step in developing effective waste management plans is to conduct an environmental audit. Environmental audits provide detailed information on the types, volumes, locations, and handling procedures of all materials that have a potential to impact the environment, and they determine whether operations are in compliance with applicable regulations. The primary objectives of environmental audits are to lower the operating, compliance, and liability costs associated with drilling and production operations. Several benefits of environmental audits are that they:

1. Determine compliance with applicable regulations.
2. Identify activities where improvements in operations are needed to minimize risk, liability, and potential environmental impacts or to lower operating costs.
3. Improve decision-making ability of facility personnel regarding environmental issues.
4. Provide an early warning device for impending problems and reduce "surprises" or repeated patterns of shortcomings in environmental performance.
5. Increase awareness among supervisors and operators of the regulatory requirements.
6. Reinforce top management's commitment to environmental protection.
7. Identify areas where environmental training is needed.
8. Establish and quantify measures for risk reduction.
9. Confirm effective communications between environmental staff and field personnel.
10. Increase confidence of management that environmental activities are a sound investment.
11. Determine how knowledgeable employees are about company policies regarding environmental issues.
12. Improve relationship with regulatory agencies and the public in regard to activities conducted by the company.

Normally, the first steps in conducting an environmental audit are to review records of the site, to interview knowledgeable people about the site and its activities, and to conduct a physical inspection of the site. If obvious problems exist or if insufficient information to evaluate the potential for future liability of the property is available, then a more detailed study involving sampling and detailed engineering analysis may be required (Curtis and Kirchof, 1993). The information that can be obtained during an audit includes records of all materials entering the area, including those produced from the wells, created in surface facilities, and brought into the area by service companies. These materials include all solids, sludges, liquids, gases, and mixtures. The volume of each of these materials and their ultimate disposition must be identified. Naturally occurring radioactive materials (NORM) generated during production must also be considered.

Environmental audits are normally conducted by a team of one to about five people, depending on the size and complexity of the facility being audited. The team members must be familiar with the full range of issues affecting the facility, including all regulatory and technical areas. They must have a knowledge of the audit process, understand all applicable regulations, have an independent viewpoint on the facility, know corporate policy, and be familiar with the history and processes used at the facility. Because audits can be viewed with hostility by those being audited, team members must also have good communication and professional skills.

The audit team develops the audit protocol, which is a detailed list of the activities that will be conducted during the audit. The protocol depends on the needs and objectives of the audit, but normally includes three steps: pre-audit activities, a field visit to the site, and some type of follow-up.

In the pre-audit activities, the goals and objectives of the audit are established, the scope, target, and subjects of the audit are selected, a schedule is developed, checklists and questionnaires are developed, materials are exchanged between the audit team and targeted facility, and all exchanged materials are reviewed.

The field visit starts with a briefing in which the purpose, authority, confidentiality arrangements, facilities, and documents are reviewed. Managers, foremen, and operations people should all be interviewed to determine their knowledge about environmental issues and company policies. The questions asked should be from a prepared questionnaire developed during the pre-audit activities. Detailed records of all permitted activities are required under most permits and should be reviewed. A tour of the facilities is then conducted to verify that operations are actually conducted according to the written plans. A facility visit should include a walk around the property line to observe possible storm runoff discharges. Following the tour, a final briefing is given. The final briefing should be a very short summary of the audit findings, with a statement that a formal written report will be forthcoming. This briefing informs local management of what senior management will be told and gives them an opportunity to prepare their response.

After the field visit, a final written report is prepared, a list of corrective measures is developed, and a follow-up visit to verify the success of the corrective measures is conducted. The report should review the program strengths, describe areas where improvement is needed, and make recommendations for corrective action to address problems. In preparing the written report, any problems can usually be grouped into one of the following three categories:

1. Activities that are in violation of permits or regulations, activities that are in violation of company policies, or policies that encourage activities to be in violation. These activities require the attention of senior management and need immediate correction.
2. Conditions that could result in a violation or in a situation that could harm the environment. Corrective actions are needed, but immediate action is not necessarily required. Corrective actions could be scheduled around site operations.
3. Local issues concerning housekeeping, storage, purchasing, or similar items. Corrective actions are recommended, but not necessarily required.

Depending on the magnitude of the problems identified and the corrective actions recommended, a follow-up visit should be conducted within 3 to 12 months after the audit.

Typical operational problems that are identified in environmental audits are compressors and engines that are installed without state permits or inventorying; gas plants and sweetening plants that are installed or modified without considering *new source performance standards* [NSPS] or *prevention of significant deterioration* [PSD] monitoring, analysis, or control technology; unapproved analytical methods that are used for determining compliance; and underground injection wells that have not received the proper permits. Typical problems with personnel training are field foremen who do not know if hazardous substances are located at their site or how to report a release to regulatory agencies if a release occurs, and engineering staff who are not well informed of environmental design requirements.

One difficulty with performing an internal environmental audit is that the U.S. Environmental Protection Agency can request a copy of the audit under certain circumstances. Unless there is a commitment by top management to correct any and all deficiencies found, the existence of the audit could leave the company and its employees open to regulatory action, including criminal penalties for willful violation of the law for not implementing corrective measures.

A number of case histories on developing environmental audits for oil and gas production facilities have been published (Guckian et al.,

1993; Tan and Hartog, 1991; Jennett, 1991; Whitehead, 1991; and Crump and O'Gorman, 1991).

5.2 WASTE MANAGEMENT PLANS

Waste management plans identify exactly how each waste stream should be managed. They ensure that appropriate engineering controls, proper waste management options, adequate recordkeeping and reporting systems, and ongoing employee training are in place. The information obtained from environmental audits can be used in developing a waste management plan.

One of the first steps in developing a waste management plan is to identify the region and scope to be covered. All materials generated within the region must be identified, quantified, and characterized. These data must include chemical toxicological, health, fire, explosive, and reactivity information. They should also include first aid procedures to be used in the event of human exposure to the material. Material Safety Data Sheets (MSDS) provide much of this information and can be obtained from chemical suppliers.

The potential for a material to migrate from a site must also be considered when determining the best way to manage it. Factors like topography, hydrology, geology, soil conditions, and the presence of sources of usable water must be evaluated. Historical rainfall and distribution data are also needed to determine soil loading conditions, to predict net evaporation rates, to determine how quickly reserves pits will dry, and to evaluate overtopping potential of open tanks and pits during storms. Other factors that must be considered are the special needs of environmentally sensitive areas such as wetlands, rain forests, arctic tundra, arctic icepack, areas where subsidence during production may occur, urban areas, historical sites, archaeological sites, protected habitats, and sites providing habitats for endangered species.

A critical factor that must also be considered in developing waste management plans is the regulatory status of each material at a site. One way to classify wastes in the United States is according to the Resource Conservation and Recovery Act (RCRA) categories of exempt and nonexempt wastes (Stilwell, 1991). Nonexempt wastes can be further classified as hazardous, nonhazardous, or special wastes, as discussed below:

- *Exempt wastes* are directly associated with drilling of an oil or gas well or generated from the exploration and production of oil and gas. Most wastes in the upstream petroleum industry fall into this classification.

- *Nonexempt nonhazardous wastes* are not directly associated with drilling and production activities for oil and gas and are not considered hazardous under RCRA, Subtitle C. Nonhazardous wastes are those that are not specifically listed as hazardous or that do not fail the hazardous criteria of reactivity, corrosivity, ignitability, or toxicity. These criteria are discussed in Appendix A.
- *Nonexempt hazardous wastes* are either listed or fail the hazardous criteria of reactivity, corrosivity, ignitability, or toxicity. An example of these wastes are solvents used to clean production equipment. Solvents generally fail the ignitability criterion.
- *Nonexempt special wastes* are covered under special statutes and regulations. Examples of wastes in this classification are asbestos, naturally occurring radioactive materials (NORM), polychlorinated biphenyls (PCBs), and pesticides.

A critical step in developing waste management plans is to identify a specific action plan for handling each and every material at all sites covered by the waste management plan. These action plans should be based on the "Hierarchy of Waste Management Principles" that were promulgated in the Pollution Prevention Act of 1990 and further defined by the U.S. Environmental Protection Agency's "Memorandum on Pollution Prevention" (Habicht, 1992). This hierarchy of waste management principles defines the preferred order for actions related to managing wastes.

The first and most important action in the waste management hierarchy is to reduce the volume of wastes generated. The next action is to reuse the wastes or materials in the wastes. Only after those actions have been completed should the remaining wastes be treated and disposed. By following this hierarchy, both the volume of waste to be disposed and the ultimate disposal cost will be minimized.

Possible actions for managing each material at a site can be identified by evaluating current practices in that area, current practices in other areas, current practices for other types of wastes, practices used by other companies or industries for similar wastes, and new practices that may be described at trade shows or in the literature. Examples of waste management actions within the Hierarchy of Waste Management Principles are given in the following section.

Once a list of possible actions has been identified, those actions need to be evaluated and prioritized and a preferred action selected. Factors to be considered include cost, practicality, future liability, regulatory status, availability of resources and facilities, company policy, and local community concerns. Actions that are unacceptable should also be identified. This evaluation can include a risk assessment study to optimize the use of the available funds (Stanley and Johnson, 1993).

A critical aspect of good waste management plans is to develop and maintain good bookkeeping practices. This bookkeeping must include a waste tracking program which identifies where the waste was generated, the date the waste was generated, the type of waste and its volume, any transportation of the waste, the disposal method and location, and the contractor employed. A waste management plan must also identify which personnel are responsible for the proper management of all wastes produced at the targeted facilities.

A number of waste management plans have been discussed in the literature (American Petroleum Institute, 1989; Benoit and Schuh, 1993; Canadian Petroleum Association, 1990; Chandler, 1991; Frampton, 1990; Greer, 1991; Huddleston et al., 1990; Jones and Woodruffe, 1991; Manning and Grannan, 1991; Sarokin et al. 1985; Stilwell, 1991; Yates, 1990). Waste management plans can also be computerized (Crump and O'Gorman, 1991; Warner, 1993; Lawrence et al., 1993). Sensitive habitats like rain forests or arctic regions may require special operating practices to protect them. These practices are discussed in Appendix B.

5.3 WASTE MANAGEMENT ACTIONS

In this section, a number of examples of waste management activities for drilling and production operations are discussed according to the hierarchy of waste management principles. These activities include ways to minimize the volume and/or toxicity of wastes generated and ways to reuse or recycle wastes. Waste treatment and waste disposal options are discussed in Chapters 6 and 7, respectively.

5.3.1 Waste Minimization

The most effective way to reduce the environmental impact associated with exploration and production of oil and gas is to minimize the total volume and/or the toxic fraction of wastes generated. The primary waste minimization activities are to make changes in how chemical inventories are managed, how operations are conducted, which materials and chemicals are used, and how equipment is operated.

The advantages of waste minimization include avoidance of waste transportation and disposal costs, elimination of expensive pollution control equipment, improved product quality, less administrative recordkeeping, lower on-site handling costs, a smaller waste storage area, reduced waste and tax obligations, improved public image, lower potential environmental impacts, and reduced future liabilities.

Unfortunately, the opportunities to significantly reduce the volume of drilling and production wastes are limited. The greatest volume of waste is produced water, which is controlled by the age and production history of the field. The volume of drilling wastes is controlled primarily by the depth and number of wells drilled. Nevertheless, many opportunities are available for minimizing wastes and have been described in the literature (Hall and Spell, 1991; Savage, 1993; Thurber, 1992; Wojtanowicz, 1993a and 1993b).

Inventory Management

One aspect of waste minimization is to carefully monitor inventories of all materials at a site. Accurate, written records of all raw and processed materials and their volumes should be kept for every stage of handling and production. The costs of each material, including disposal, should also be recorded.

Better management of materials inventories provides significant environmental benefits. It allows a material balance to be conducted on all materials at all stages of usage. A detailed material balance can help identify where unwanted losses and waste may be occurring. From a better understanding of actual needs of different materials, the volumes of chemicals purchased may be reduced. Keeping excess chemicals in stock increases both the cost and the chance of spillage or leakage. Accurate records also allow chemicals to be rotated so that their shelf life does not expire before they are used. If large volumes of a chemical are needed, it can be purchased in bulk to reduce the number of containers requiring disposal. Accurate records can be used to determine whether the volume of chemicals purchased are proportional to their use and whether purchase restrictions are needed.

Improved Operations

Another important method for minimizing the amount of potentially toxic wastes generated is to change the operating procedures at the various sites. Many changes can be made to improve operations at a relatively low cost, particularly if planned in advance.

New sites can be constructed to minimize environmental risks. Access roads can be sited to minimize any disruption. Sites should be kept as small as possible and should be designed so that natural drainage features will divert rainwater around the site, in particular, away from the rig and reserves pits. The soil type should be evaluated to determine if it is suitable for constructing site facilities such as buildings, drill pads, ponds, levees, or production tanks. Dikes and catchment basins should be constructed around all storage tanks and loading areas to contain any leaks and spills. If a site is suspected of being contaminated by any previous activity, a detailed site assessment should be conducted to characterize any contamination before any new activity commences.

All operations should be carefully planned in advance to minimize the use of materials. Materials storage, handling, and transportation procedures should be reviewed to minimize losses. Only the required amount of chemicals and equipment should be available at the site.

A very important step in improving operations is to keep different types of wastes segregated. Waste streams should never be mixed. Because the toxicities and regulations vary for different wastes, keeping the waste streams segregated allows the best disposal options to be selected for each waste. This minimizes the volume of toxic wastes that must be handled under the most stringent and expensive regulations. For example, hazardous and nonhazardous wastes should never be mixed. Municipal or commercial wastes should be kept separate from other site wastes. Soil contaminated with hazardous and/ or commercial wastes should be kept separate from soil contaminated with other wastes. Sites should be designed to keep unwanted materials from entering the fluid system and reserves pit during drilling. This unwanted material includes rig wash, pump lubrication water, drill pipe handling, and stormwater runoff. Levees or ditches can be used around a site to divert stormwater or contain any spills.

Good housekeeping practices must be observed at all sites. Trash containers should be provided at drill sites and production facilities to discourage disposal of refuse, paper, and other household trash in pits. These household wastes should be collected and stored separately for off-site disposal. Sanitary wastes should be collected and treated to satisfy state and local effluent requirements using septic systems, portable commercial containers, shipment to municipal sewage facilities, or disposal at municipal solid waste sites. Leaks and spills from all equipment should be eliminated. Liners, drip pans, or basins can be used to collect any potential spillage from equipment. Equipment that is leaking should be repaired or replaced. Drilling rigs should be washed at a site only if absolutely needed and only with recycled pit or sump water, not with fresh water.

Optimized drilling operations provide a significant opportunity for minimizing wastes. Because the total volume of drilling wastes is controlled primarily by the hole size and well depth, the smallest diameter hole should be drilled to minimize the volume of cuttings generated and drilling mud used. The needs for future recovery activities, including possible multiple tubing strings for improved recovery operations, must be considered when determining the hole size. Intermediate casing strings can be used to isolate problem zones, e.g., salt, high pressure, or reactive shales, and minimize the volume of specialized drilling mud needed to drill below those zones. Hole washout can be minimized during drilling by limiting the recirculation rate of drilling fluids such that the annular velocity to lift cuttings is not excessive. During drilling, the surge and swab pressures in the wellbore should be minimized by limiting rapid pipe movement to maintain a good mudcake and prevent further hole enlargement.

A number of drilling mud systems are available. Closed-loop systems with good solids control and separation equipment can be used to minimize the volume of drilling wastes. In these systems, covered steel tanks are used instead of digging reserves pits in the ground. Advanced solids separation and dewatering equipment must be used, however. Drilling fluid systems and fluids should be designed to minimize drilled-solids degradation and reduction of particle size. The bottom of the mousehole should be cemented to prevent drilling fluids from leaching into groundwater when the kelly is in place.

Reserves pits can be constructed to minimize the volume of wastes. They should cover a limited area to control the amount of rainfall entering them, but they should also have sufficient capacity so they do not overflow during heavy rains. In many areas, pit liners are required if saltwater- or oil-based muds are used. Liners should be considered in sensitive areas, even if not required by current regulations. Netting may be required over pits to prevent birds from landing in them, particularly if the pits contain floating hydrocarbons.

Because many different fluids are found at most drill sites, a managed pit system with multiple pits can be used to keep the different fluids separated (Hall et al., 1991; Pontiff et al., 1990). For example, one pit could be used for mud reserves, one for cleaned cuttings, one for skimmed oil, one for kicks, and one for stormwater runoff. If different mud types are used for different parts of the well, e.g., when drilling through an overpressured layer, a salt dome, or other very sensitive formation, a separate pit can be used for each mud type. One pit can be used for drilling the top section of the well where native materials can be used with minimal additives and one pit can be used for drilling through the productive horizon, particularly for horizontal wells. These systems minimize the total volume of materials having the greatest potential for environmental impact. With a managed pit system, different materials can be disposed of in the best way, minimizing the volume of materials that must be disposed of at the highest cost and reducing future liability.

Preplanning is important when developing a mud for each well. When selecting a mud, detailed questions should be asked to the sales representative about exactly what the various additives will do and whether they are actually necessary for a particular well. Mud additives can be pilot tested in small volumes to ensure they behave as claimed. Inhibitive mud should be used to minimize hole enlargement during drilling from the hydration of shales. Mud density changes should be avoided because these normally require discarding some of the mud and reformulating the remainder; this leads to an increase in the total volume of mud that is used.

A number of operational changes during production can also be implemented to minimize the total volume of waste generated. Routine inspection and/or pressure testing of all tanks, vessels, gathering lines, and flow lines should be scheduled. Routine inspection and/or automatic pumps should be installed in all sumps.

Unfortunately, the largest volume of production waste is produced water and little can be done to minimize its production. In some formations, coning of water can be minimized by dually completing a well in both the water and oil zone. This can limit water coning and reduce the amount of water produced with the oil. The water produced from the water zone must also be disposed of (reinjection), but it should contain essentially no oil (Wojtanowicz, 1991). Polymers, gels, or cement can be used to plug water production zones if they are separate from oil producing zones.

Since many drilling and production operations are conducted by contractors, they should be carefully reviewed and selected. Contractors should have a good environmental track record. When conducting the bidding process for selecting equipment to be used, a visual inspection of the equipment is advised to determine its general condition, particularly drilling rigs. Contractors should have properly functioning equipment, with drip pans and splash guards.

Any contracts should specify activities that are prohibited while the contractor is on site. Such activities can include unnecessary rig washing, painting of the contractor's equipment, or changing lube oil during downtime. This will minimize the probability that excess water, painting wastes, or used oil gets dumped into reserves pits. An environmental activity review should be conducted with all contractor crews just prior to the start of activities. This review should include waste handling and minimization procedures.

Materials Substitution

Another important method for minimizing the amount of potentially toxic wastes generated is to use less toxic materials for the various operational processes. A number of studies of material substitutions have been presented (Derkics and Souders, 1993; Freidheim and Shinnie, 1991; Peresich et al., 1991; Savage, 1993; Thurber, 1992; Wojtanowicz, 1991).

Drilling muds represent a significant opportunity for toxic waste reduction by materials substitution. When substituting materials, however, it is important to ensure that the substituted materials yield a drilling mud that still has acceptable properties.

One of the best opportunities for materials substitution is in wells where oil-based muds are needed. Two alternatives to the use of diesel oil as a base fluid are being studied: using a less toxic oil-based mud and using a water-based mud with an improved additives package. These alternative mud systems, however, are considerably more expensive than traditional muds. Unfortunately, the use and discharge of these new drilling muds may still be prohibited, even though they provide significantly improved environmental protection. Many of the regulations covering the discharge of drilling muds were established before the development of these alternative mud systems and have not been changed to reflect these new technologies.

One way to lower the toxicity of diesel oil muds is to increase the amount of water in the mud emulsion. This will reduce the amount of oil that is available to be retained on cuttings. Water contents in new mud formulations have been reported to be as high as 65% (Friedheim and Shinnie, 1991). Traditional muds have water contents typically around 10% or less. Another way to lower the toxicity of oil-based muds is to use a less toxic base oil. Mineral oils having a low concentration of aromatic hydrocarbons have been successfully used (Jacques et al., 1992), as have esters, ethers, and polyalphaolefins (Peresich et al., 1991; Candler et al., 1993). Cationic surfactants can also be added to the mud to reduce the amount of oil trapped on cuttings (Friedheim and Shinnie, 1991).

A variety of new water-based muds are being developed as possible substitutes for oil-based muds. The additives for these muds have included various low-toxicity polymers and glycols (Bland, 1992; Bleier et al., 1993; Enright et al., 1991; Reid et al., 1993). Substitutions can also be made with the additives used in water-based muds. For example, dolomite can be used instead of barite as a weighting agent. Additives made from water-soluble combinations of silicon, phosphorus, aluminum, and boron can replace some conventional additives (Zakharov and Konovalov, 1992). New pipe dopes are being developed that do not contain heavy metals; these new pipe dopes have included micron-sized alumina-ceramic beads in a lithium grease.

Drilling muds can be reformulated to improve shale stability. This will reduce wellbore washouts, minimize the degradation of solids (the breaking into smaller, harder-to-separate particles), reduce the amount of material brought to the surface to be handled, and lower the mud volume requirement of the well (Alford, 1991; Thurber, 1992). Potassium acetate or potassium carbonate can be used instead of potassium chloride for shale stability problems to minimize the chloride content of the drilling mud (Gillenwater and Ray, 1989). Other mud additives and suggested substitute materials are given in Tables 5-1 and 5-2.

A variety of opportunities are available during production operations to substitute less toxic materials for more toxic, traditional materials. For

Table 5-1

Substitute Materials for Drilling Fluid Additives

Additive	Toxic Component	Use	Substitute Material
Chrome lignosulfonate/ lignite	Chromium	Deflocculant	Polyacrylate and/or polyacrylamide polymers
Sulfomethylated tannin/dichromate	Chromium	Deflocculant	Polyacrylate and/or polyacrylamide polymers
Sodium chromate	Chromium	Corrosion control	Sulfites, phosphonates, and amines
Zinc chromate	Chromium	H_2S control	Nonchromium H_2S scavengers
Pentachlorophenol	Pentachlorophenol	Biocide	Isothiazolins, carbamates, and gluteraldehydes
Paraformaldehyde	Formaldehyde	Biocide	Isothiazolins, carbamates, and gluteraldehydes
Arsenic	Arsenic	Biocide	Isothiazolins, carbamates, and gluteraldehydes
Lead-based pipe dope	Lead	Pipe thread sealant/lubricant	Unleaded pipe dope
Barite	Cadmium/mercury/ barium/lead	Mud densifier	Chose barite from sources low in cadmium, mercury, and lead.

Source: after Thurber, 1992.
Copyright SPE, with permission.

Table 5-2
Additional Substitute Materials for Drilling Fluid Additives

Hazardous Item	Substitute
Pipe dope compounds: lead, zinc, copper, and cadmium	Lithium-based grease with microsphere ceramic balls
Oils and greases: aromatics, sulfur	White oils manufactured from highly refined mineral oils approved for use in the food industry
Cleaning solvents: varsol, freon, MEK, phosphate soaps	Citrus-based solvents, high pressure hot water, jet washers, closed-loop recycling

Source: after Page and Chilton, 1991.
Copyright SPE, with permission.

example, organic cations can be used as a low salt concentration, temporary clay stabilizer in well service fluids (Himes, 1991; Himes et al., 1990). Zinc, sulfite, or organic phosphate corrosion inhibitors can be used instead of chromate inhibitors. Pentachlorophenols and formaldehyde-releasing biocides can be replaced with isothiazoline or amines. Petroleum- and alcohol-based defoamers can be replaced with polyglycols.

Opportunities for materials substitution are also available during related site operations. For example, less toxic detergents can be used to wash rigs. A better solution, however, is for contractors to install closed-loop washwater systems for washing rigs at their own sites rather than at the wellhead (Whitney and Greer, 1991). Whenever possible, unleaded water-based paints and nonsolvent paint removers, cleaners, and degreasers can be used. Disposable brushes can be used to eliminate the need for paint thinners and solvents, although the brushes must then be disposed of. Water-based dyes can be used instead of trichloroethane-based penetrants when inspecting pipes for cracks. Substitutes can be used for halon gases in fire suppressants.

Equipment Modifications

Another important method for minimizing the volume of potentially toxic wastes generated is to ensure that all equipment is properly operated and maintained. Inefficient equipment should be replaced with newer, more efficient equipment.

One of the first steps to be taken is to eliminate all leaks and spills from equipment. Drip pans can be used beneath the drilling rig floor to catch all water or mud drained from it. Flexible hoses can be used to drain water to or from the cellar. Leaking stuffing box seals should be replaced or new stuffing boxes installed. Fugitive emissions from leaking valves, flanges, and such fittings can be minimized by replacing leaking equipment.

If the interval between lube oil changes on diesel engines is lengthened, the volume of waste lube oil can be reduced. The interval recommended by manufacturers is normally based on "worst case" conditions operations. By monitoring the quality of the lube oil over time and using a higher quality lube oil, it may be possible to increase the time between changes without any loss of engine protection. This could significantly reduce the total amount of lube oil used (Reller, 1993).

Internal combustion engines should be properly tuned and the proper fuel should be used. The emission of partially burned hydrocarbons can be minimized by control of the fuel/air ratio during combustion. The formation of SO_x during combustion can be minimized by using a low sulfur fuel such as natural gas.

If the volume of waste generated cannot be sufficiently reduced with the existing equipment, newer equipment should be installed. Important environmental features of newer equipment should be how easy they are to monitor and clean up, as well as how they facilitate waste recovery and recycling. New equipment should have modern emission controls. In some cases, equipment with automated process controls can be installed to ensure optimal operations.

Automatic shutoff nozzles and low-volume, high-pressure nozzles should be installed on all hoses on the rig floor and wash racks to minimize wastewater. Water meters should be installed on all fresh-water sources to monitor and control water usage. Rig wash should be limited to only the minimum needed for safety, not for esthetics.

More efficient separations equipment should be used to separate solids, hydrocarbons, and water. Newer shale shakers can be used that are better at filtering out small solids than older equipment. Low shear pumps should be used for produced water to prevent hydrocarbon droplets from decreasing in size, because small droplets are more difficult to remove. Improved backwash equipment and better procedures can be used to extend filter life.

The most important way to reduce the emission of volatile hydro-carbons at production facilities is to install vapor recovery systems. Casing vapor recovery systems should be in thermal recovery operations to collect casing gases. Recovery units can be installed to collect glycol reboiler vapors (Choi and Spisak, 1993; Schievelbein, 1993). Mercury manometers along gas flow lines can be replaced with electronic, digital flow meters

5.3.2 Material Reuse

Many of the materials in drilling and production waste streams can be used more than once. If materials are intended for future use, they are not wastes. The following materials have a potential for reuse: acids, amines, antifreeze, batteries, catalysts, caustics, coolants, gases, glycols, metals, oils, plastics, solvents, water, wax, and some hazardous wastes.

Water has a considerable potential for reuse. For example, water from reserves pits can be used to wash shale shakers and other solids control equipment during drilling. Reserves pit water should also be used as makeup water for drilling mud as much as possible. Water from mud can be cleaned and used as rig washwater. Rig washwater can be collected and reused, particularly at contractor facilities. Lubrication and cooling water used by pumps can also be recycled. Water obtained from dewatering a reserves pit could be treated and used at another site, particularly in arid areas. Produced water, after treatment, can be reinjected for pressure maintenance during water floods or for steam injection in heavy oil recovery.

Material reuse can be facilitated by installing equipment that allows reuse. For example, closed-loop systems can be installed so that solvents and other materials can be collected and reused in plant processes. Reusable lube oil filters can be installed in some applications instead of throwaway filters. Flared natural gas can be reinjected for pressure control, or an alternate use for it can be found. Flaring should be restricted to emergency conditions only.

Many drilling and production wastes could be used at other sites or be returned to the vendor. For example, reconditioned drilling mud could be reused for other wells, either by the operating company or by the vendor. Waste mud from one well can be used for plugging or spudding other wells. Some used chemical containers can be returned to the vendor for refilling. Oily rags can be cleaned and reused. Used drilling mud can also be used to make cement. Waste acids can be used to neutralize caustic wastes, and vice versa.

Many wastes can be used as feedstock by other companies. Materials exchanges are available in numerous locations to assist companies in finding other companies that may be interested in obtaining wastes. These exchanges should be contacted to see exactly what materials can be recycled in each area. Care should be taken, however, that the recycler is reputable and in compliance with all regulations. Transfer of a waste to a waste exchange does not necessarily relieve the waste generator of future liability for what the waste exchange does with the waste. A list of some of the major waste exchanges in the United States is given in Appendix C (Quan, 1989).

In some cases, only part of a particular waste stream contains valuable materials that can be reused. It may be possible to recover or reclaim the valuable materials, reducing the net volume of waste. For example, crude oil tank bottoms, oily sludges, and emulsions can be treated to recover their hydrocarbons. Oily materials can also be burned for their energy content. Gravel and cuttings can be washed and used in construction of roads and other sites.

Companies can take proactive action to assist employees in finding suitable opportunities for recycling. For example, funds generated from recycling can be placed in an employee fund for use at employee discretion to encourage recycling. Emphasis can also be placed on purchasing recycled goods to increase the market for them.

5.3.3 Treatment and Disposal

Wastes that cannot be eliminated must be treated and disposed. Treatment is used to reduce the volume and/or toxicity of wastes and/or put it in a form suitable for final disposal. A number of treatment and disposal options are available for the wastes generated in the petroleum industry. These options are discussed in Chapters 6 and 7, respectively.

5.4 CERTIFICATION OF DISPOSAL PROCESSES

One option for waste management is to ship wastes to an off-site, commercial waste disposal facility. Paying a disposal facility to take wastes, however, does not necessarily remove liability for what subsequently happens to those wastes. Because the company that generated the wastes normally retains liability, great care should be exercised in selecting and using commercial waste disposal facilities. One way to minimize the risk of liability after custody of the waste has been transferred is to develop a formal certification process (Steingraber et al., 1990).

The first step in the certification of a waste disposal facility is to gather as much information about the facility as possible. This information includes institutional information, which includes its conformance record for existing rules and regulations, its operational and physical capabilities, and the geologic and hydrologic conditions at the site. A detailed site visit should also be conducted. A set of criteria for deciding whether a facility is acceptable or not must also be developed. If a facility has been certified to be acceptable and wastes are shipped to it, the facility should be reevaluated on a regular basis.

Part of the certification process for off-site disposal of wastes is an evaluation of how the wastes are transported to the facility. Reputable haulers that have all necessary permits for waste transportation must be selected. Manifests of all materials shipped are also required to maintain a paper trail on the disposition of the wastes.

5.5 CONTINGENCY PLANS

Contingency plans are needed to prepare a facility to minimize the impact of any foreseeable emergency. Contingency plans for environmental protection outline the response of all personnel to an accidental release of materials that can impact the environment. These plans describe ways to eliminate the source of the release, to assess the character, amount, and extent of the release, to identify ways of containing the release so any impacts are minimized, to recover all lost or contaminated materials, and to notify relevant regulatory authorities. Contingency plans must carefully and completely document the response of all personnel in the event of an emergency (Tomlin and Snider, 1994).

Contingency plans supplement, but do not replace, waste management plans. They provide a framework to prepare for and handle all significant risk scenarios. Like all waste management plans, contingency plans should be in writing. A contingency plan must be accepted by those with authority to approve or deny its implementation. Prior review and approval of such a plan by regulatory agencies will help in obtaining final approval when an emergency occurs. Prior review and approval, however, does not necessarily mean final approval will be given to implement the plan in an emergency.

Contingency plans can be developed through the following steps (Geddes et al., 1990).

Step 1: Identify potential emergencies and complications. All possible emergencies and complicating factors are to be identified. All scales of emergencies should be considered.

Step 2: Identify risks and consequences. In this step, the potential impacts of emergencies on human life, wildlife, and the environment are determined.

Step 3: Identify resources and capabilities. This step requires a detailed assessment of all resources available to meet any emergencies. Resources to be evaluated include personnel, equipment, supplies, and funds.

Step 4: Determine and define roles and responsibilities. The actions of all personnel during an emergency, including field hands, management, and regulatory agencies, are outlined. Communication channels are also clearly explained.

Step 5: Determine response actions. A realistic, detailed plan of action is outlined for each potential emergency. It should include the estimated timing of equipment arrival, operations of the equipment (including operating personnel), and decision-making priorities.

Step 6: Write and implement the plan. The plan should be written in easy-to-understand language and should be user-friendly. It should allow for updates and modifications. It should be considered a "live" plan, i.e., it should be changed as needs and experience dictate. It is important for the surrounding community to be informed of the plan and to have input into it as it is developed and modified.

A number of contingency plans are required in the United States by federal regulations. For example, the Clean Water Act requires that spill prevention control and countermeasure (SPCC) plans be developed to minimize the risk of accidental discharge of oil. The Oil Pollution Act requires a response plan for actions following the accidental release of oil. The Occupational Safety and Health Administration's "Hazardous Waste Operations and Emergency Response" (HASWOPER) requires a plan to protect worker health and safety in cleanup operations at waste sites.

5.6 EMPLOYEE TRAINING

For any environmental protection plan to be effective, it must be understood and accepted by those who must implement it. Best results are normally obtained by establishing a formal training program for all employees who make decisions that can impact the environment. Once developed, environmental protection plans will serve as handy guides for all the people to use in making the best decisions regarding wastes.

A critical step in the effective implementation of the environmental protection plan is to identify the people involved with the actual decisions impacting the environment and to effectively communicate the plan to them. Employees need easy access to information on approved methods for handling, treating, and disposing of different waste streams, as well as applicable regulations. In many cases, the first- and second-line production and drilling supervisors will be the primary users of the plan. It is important that they are provided clear, concise directives on what is required of their operations. These directives should include appropriate background information.

Because different operations within a company have different needs, it may be necessary to have a series of separate plans and training programs to meet those diverse needs. For example, managers, engineers, field foremen, and pumpers need different information to complete their tasks. A one-page summary can be prepared for use in the field that gives a quick reference on how each waste is to be handled. This page can be incorporated into a plant operator's or pumper's field book and posted on bulletin boards. A detailed manual giving more complete information should be prepared and kept as a reference manual in various offices. One such reference manual is available from the Canadian Petroleum Association (1990).

When the plan is written, it is important that it be composed so the field people can easily understand it, i.e., it must be user-friendly. To ensure readability, the plan and the manuals should be reviewed by field personnel before being adopted. Compliance with the plan by field personnel will be enhanced by having them contribute to it during its formulation and development. The plan must also be reviewed and approved by the management and legal staff of the company.

Once the plan has been written and approved, an effective training program should be implemented to educate company personnel on its contents. This program should include adding environmental issues to job descriptions and including environmental compliance during job performance evaluation for promotion and merit salary increases.

REFERENCES

Alford, S. E., "North Sea Field Application of an Environmentally Responsible Water-Base Shale Stabilizing System," paper SPE/IADC 21936 presented at the Society of Petroleum Engineers 1991 Drilling Conference, Amsterdam, The Netherlands, March 11–14, 1991.

American Petroleum Institute, "API Environmental Guidance Document: Onshore Solid Waste Management in Exploration and Production Operations," Washington, D.C., Jan. 1989.

Benoit, J. R. and Shuh, M. G., "Waste Minimization at Sour Gas Facilities," paper SPE 26011 presented at the Society of Petroleum Engineers/Environmental Protection Agency's Exploration and Production Environmental Conference, San Antonio, TX, March 7–10, 1993.

Bland, R., "Water-Based Glycol Systems Acceptable Substitute for Oil-Based Muds," *Oil and Gas J.,* June 29, 1992, pp. 54–59.

Bleier, R., Leuterman, A. J. J., and Stark, C., "Drilling Fluids: Making Peace with the Environment," *J. Pet. Tech.,* Jan. 1993, pp. 6–10.

Canadian Petroleum Association, "Production Waste Management Handbook for the Alberta Petroleum Industry," Dec. 1990.

Candler, J. E., Rushing, J. H., and Leuterman, A. J. J., "Synthetic-Based Mud Systems Offer Environmental Benefits Over Traditional Mud Systems," paper SPE 25993 presented at the Society of Petroleum Engineers/Environmental Protection Agency's Exploration and Production Environmental Conference, San Antonio, TX, March 7–10, 1993.

Chandler, J., "Environmental Rules Make Scrutiny of Rig Work Practices a Must," *Oil and Gas J.,* April 22, 1991, pp. 83–85.

Choi, M. S. and Spisak, C. D., "Aromatic Recovery Unit (ARU): A Process Enhancement for Glycol Dehydrators," paper SPE 25953 presented at the Society of Petroleum Engineers/Environmental Protection Agency's Exploration and Production Environmental Conference, San Antonio, TX, March 7–10, 1993.

Crump, J. J. and O'Gorman, T. P., "A Task Management System for Compliance with Health, Safety, and Environmental Regulations," paper SPE 22292 presented at the Society of Petroleum Engineers Sixth Petroleum Computer Conference, Dallas, TX, June 17–20, 1991.

Curtis, B. W., II., and Kirchof, C. E., Jr., "Purchase/Sale of Property: The Black Hole of Corporate Liability, Ways to Minimize Risk," paper SPE 25957 presented at the Society of Petroleum Engineers/Environmental Protection Agency's Exploration and Production Environmental Conference, San Antonio, TX, March 7–10, 1993.

Derkics, D. L. and Souders, S. H., "Pollution Prevention and Waste Minimization Opportunities for Exploration and Production Operations," paper SPE 25934 presented at the Society of Petroleum Engineers/Environmental Protection Agency's Exploration and Production Environmental Conference, San Antonio, TX, March 7–10, 1993.

Enright, D. P., Dye, W. M., and Smith, F. M., "An Environmentally Safe Water-Based Alternative to Oil Muds," *SPE Drilling Engineering,* March 1992, pp. 15–19.

Friedheim, J. E. and Shinnie, J. R., "New Oil-Base Mud Additive Reduces Oil Discharged on Cuttings," paper SPE/IADC 21941 presented at the Society of Petroleum Engineers 1991 Drilling Conference, Amsterdam, The Netherlands, March 11–14, 1991.

Frampton, M. J., "Waste Management Decision Making Procedure at Prudhoe Bay, Alaska," Proceedings of the U.S. Environmental Protection Agency's First International Symposium on Oil and Gas Exploration and Production Waste Management Practices, New Orleans, LA, Sept. 10–13, 1990, pp. 1071–1080.

Geddes, R. L. Fraser, I. M., and Berezuk, Z. L., "Contingency Plans for Beaufort Sea Drilling into the 1990s," paper CIM/SPE-90-140, presented at the Canadian Institute of Mining International Technical Meeting, Calgary, Alberta, June 10–13, 1990.

Gillenwater, K. E. and Ray, C. R., "Potassium Acetate Adds Flexibility to Drilling Muds," *Oil and Gas J.,* March 20, 1989, pp. 99–102.

Greer, C. R., "Managing Environmental Compliance for Field Facilities," paper SPE 23510 presented at the Society of Petroleum Engineers First International Conference on Health, Safety, and Environment, The Hague, Netherlands, Nov. 10–14, 1991.

Grogan, W. C., "The Use of Environmental Assessments in the Brae Field Development," paper SPE 23328 presented at the Society of Petroleum Engineers First International Conference on Health, Safety, and Environment, The Hague, Netherlands, Nov. 10–14, 1991.

Guckian, W. M., Hurst, K. G., Kerns, B. K., Moore, D. W., Siblo, J. T., and Thompson, R. D., "Initiating an Audit Program: A Case History," paper

SPE 25955 presented at the Society of Petroleum Engineers/Environmental Protection Agency's Exploration and Production Environmental Conference, San Antonio, TX, March 7–10, 1993.

Hall, C. R., Ramos, A. B., Oliver, R. D., and Favor, J., "The Use of a Managed Reserve Pit System to Minimize Environmental Costs in the Pearsall Field," paper SPE 22882 presented at the Society of Petroleum Engineers 66th Annual Technical Conference and Exhibition, Dallas, TX, Oct. 6–9, 1991.

Hall, C. R. and Spell, R. A., "Waste Minimization Program can Reduce Drilling Costs," *Oil and Gas J.,* July 1, 1991, pp. 43–46.

Habicht, F. H., "EPA Memorandum on Pollution Prevention" (May 28, 1992), U.S. Environmental Protection Agency, Bureau of National Affairs, Inc. Washington, D.C., July 1992.

Himes, R. E., "Environmentally Safe Temporary Clay Stabilizer for Use in Well Service Fluids," *Advances in Filtration and Separation Technology, Vol. 3, Pollution Control Technology for Oil and Gas Drilling and Production Operations,* American Filtration Society. Houston: Gulf Publishing Co., 1991, pp. 124–139.

Himes, R. E., Parker, M. A., and Schmeizl, E. G., "Environmentally Safe Temporary Clay Stabilizer for Use in Well Service Fluids," paper CIM/SPE 90-142 presented at the Canadian Institute of Mining International Technical Meeting, Calgary, June 10–13, 1990.

Huddleston, R. D., Ross, W. A., Benoit, J. R., "The Development of a Waste Management System for the Up-Stream, On-Shore Oil and Gas Industry in Western Canada," Proceedings of the Society of Petroleum Engineers/Environmental Protection Agency's First International Symposium on Oil and Gas Exploration and Production Waste Management Practices, New Orleans, LA, Sept. 10–13, 1990, pp. 227–242.

Jacques, D. F., Newman, H. E., Jr., and Turnbull, W. B., "A Comparison of Field Drilling Experience with Low-Viscosity Mineral Oil and Diesel Muds," paper IADC/SPE 23881 presented at the Society of Petroleum Engineers 1992 IADC/SPE Drilling Conference, New Orleans, LA, Feb. 18–21, 1992.

Jennett, L. E., "Environmental Audits for Oilfield Service Districts Methodology, Findings, and Recommendations," paper SPE 23489 presented at the Society of Petroleum Engineers First International Conference on Health, Safety, and Environment, The Hague, Netherlands, Nov. 10–14, 1991.

Jones, M. G. and Woodruffe, J. D., "Environmentally Sustainable Economic Development: E&P Planning in the 1990s," paper SPE 23341 presented at the Society of Petroleum Engineers First International Conference on Health, Safety, and Environment, The Hague, Netherlands, Nov. 10–14, 1991.

Lawrence, A. W., Miller, J. A., Miller, D. L., and Linz, D. G., "An Evaluation of Produced Water Management Options in the Natural Gas Production Industry," paper SPE 26004 presented at the Society of Petroleum Engineers/ Environmental Protection Agency's Exploration and Production Environmental Conference, San Antonio, TX, March 7–10, 1993.

Manning, L. and Grannan, S. E., "Laboratory Waste-Management Programs for Research and Field-Support Operations in the Oilfield Servicing Industry," paper SPE 23376 presented at the Society of Petroleum Engineers First International Conference on Health, Safety, and Environment, The Hague, Netherlands, Nov. 10–14, 1991.

Page, W. B. and Chilton, C., "An Integrated Approach to Waste Management," paper SPE 23365 presented at the Society of Petroleum Engineers First International Conference on Health, Safety, and Environment, The Hague, Netherlands, Nov. 10–14, 1991.

Peresich, R. L., Burrell, B. R., and Prentice, G. M. "Development and Field Trial of a Biodegradable Invert Emulsion Fluid," paper SPE/IADC 21935 presented at the Society of Petroleum Engineers 1991 Drilling Conference, Amsterdam, The Netherlands, March 11–14, 1991.

Pontiff, D., Sammons, J., Hall, C. R., and Spell, R. A., "Theory, Design, and Operation of an Environmentally Managed Pit System," Proceedings of the U.S. Environmental Protection Agency's First International Symposium on Oil and Gas Exploration and Production Waste Management Practices, New Orleans, LA, Sept. 10–13, 1990, pp. 977–986.

Quan, B., "Waste Exchanges," in *Standard Handbook of Hazardous Waste Treatment and Disposal,* H. M. Freeman (editor). New York: McGraw-Hill Book Company, 1989.

Reid, P. I., Elliott, G. P., Minton, R. C., Chambers, B. D., and Burt, D. A., "Reduced Environmental Impact and Improved Drilling Performance with Water-Based Muds Containing Glycols," paper SPE 25989 presented at the Society of Petroleum Engineers/Environmental Protection Agency's Exploration and Production Environmental Conference, San Antonio, TX, March 7–10, 1993.

Reller, C. E., "Waste Oil Reduction for Diesel Engines," paper SPE 26012 presented at the Society of Petroleum Engineers/Environmental Protection Agency's Exploration and Production Environmental Conference, San Antonio, TX, March 7–10, 1993.

Sarokin, D. J., Muir, W. R., Miller, C. G., and Sperber, S. R., "Cutting Chemical Wastes—What 29 Organic Chemical Plants Are Doing to Reduce Hazardous Wastes," INFORM, Inc., New York, 1985.

Savage, L. L., "Even if You're on the Right Track, You'll Get Run Over If You Just Sit There: Source Reduction and Recycling in the Oil Field," paper SPE 26009 presented at the Society of Petroleum Engineers/Environmental

Protection Agency's Exploration and Production Environmental Conference, San Antonio, TX, March 7–10, 1993.

Schievelbein, V. H., "Hydrocarbon Recovery from Glycol Reboiler Vapor with Glycol-Cooled Condensers," paper SPE 25949 presented at the Society of Petroleum Engineers/Environmental Protection Agency's Exploration and Production Environmental Conference, San Antonio, TX, March 7–10, 1993.

Stanley, C. C. and Johnson, P. C., "An Exposure/Risk-Based Corrective Action Approach for Petroleum-Contaminated Sites," paper SPE 25982 presented at the Society of Petroleum Engineers/Environmental Protection Agency's Exploration and Production Environmental Conference, San Antonio, TX, March 7–10, 1993.

Steingraber, W. A., Schultz, F., and Steimle, S., "Mobil Waste Management Certification System," Proceedings of the U.S. Environmental Protection Agency's First International Symposium on Oil and Gas Exploration and Production Waste Management Practices, New Orleans, LA, Sept. 10–13, 1990, pp. 599–610.

Stilwell, C. T., "Area Waste-Management Plans for Drilling and Production Operations," *J. Pet. Tech.,* Jan. 1991, pp. 67–71.

Tan, G. T. and Hartog, J. J., "Environmental Auditing in Exploration and Production Companies: A Tool for Improving Environmental Performance," paper SPE 23390 presented at the Society of Petroleum Engineers First International Conference on Health, Safety, and Environment, The Hague, Netherlands, Nov. 10–14, 1991.

Thurber, N. E., "Waste Minimization for Land-Based Drilling Operations," *J. Pet. Tech.,* May 1992, pp. 542–547.

Tomin, B. K. and Snider, R. S., "Writing Plant Emergency Manuals," Proceedings of the 1994 Petro-Safe Conference, Houston, TX, 1994.

Warner, J. W., "Environmental Data Management System," paper SPE 26363 presented at the Society of Petroleum Engineers 68th Annual Technical Conference and Exhibition, Houston, TX, Oct. 3–6, 1993.

Whitehead, A., "Environmental Auditing in European Operations," paper SPE 23391 presented at the Society of Petroleum Engineers First International Conference on Health, Safety, and Environment, The Hague, Netherlands, Nov. 10–14, 1991.

Whitney, P. M. and Greer, C. R., "Evaluation and Comparison of Closed-Loop Wash-Water Systems," paper SPE 23378 presented at the Society of Petroleum Engineers First International Conference on Health, Safety, and Environment, The Hague, Netherlands, Nov. 10–14, 1991.

Wojtanowicz, A. K., "Environmental Control Potential of Drilling Engineering: An Overview of Existing Technologies," paper SPE/IADC 21954 presented at the Society of Petroleum Engineers 1991 Drilling Conference, Amsterdam, The Netherlands, March 11–14, 1991.

Wojtanowicz, A. K., "'Dry' Drilling Location—An Ultimate Source Reduction Challenge: Theory, Design, and Economics," paper SPE 26013 presented at the Society of Petroleum Engineers/Environmental Protection Agency's Exploration and Production Environmental Conference, San Antonio, TX, March 7–10, 1993a.

Wojtanowicz, A. K., "Oilfield Environmental Control Technology: A Synopsis," *J. Pet. Tech.*, Feb., 1993b, pp. 166–172.

Yates, H., "Onshore Solid Waste Management in Exploration and Production Operations," Proceedings of the U.S. Environmental Protection Agency's First International Symposium on Oil and Gas Exploration and Production Waste Management Practices, New Orleans, LA, Sept. 10–13, 1990, pp. 703–714.

Zakharov, A. P. and Konovalov, E. A., "Silicon-Based Additives Improve Mud Rheology," *Oil and Gas J.*, Aug. 10, 1992, pp. 61–64.

CHAPTER 6

Waste Treatment Methods

During drilling and production activities, many wastes are generated that must be treated. The purpose of waste treatment is to lower the potential hazards associated with a waste by reducing its toxicity, minimizing its volume, and/or altering its state so that it is suitable for a particular disposal option. For many wastes, treatment is required prior to final disposal. A variety of treatment methods are available for most wastes, but not all methods can be used on all waste streams. The different treatment methods vary considerably in effectiveness and cost.

Most waste treatment processes involve separating a waste stream into its individual components, e.g., removing dissolved or suspended hydrocarbons and solids from water or removing hydrocarbons from solids. In many cases, a series of methods may be needed to obtain the desired treatment levels (Schmidt and Jaeger, 1990).

This chapter describes a variety of processes to treat water and solids for subsequent reuse or disposal. It also describes treatment processes for various air pollutants. More detailed discussions of treatment and disposal methods are available in the literature (Freeman, 1989; Tchobanoglous and Burton, 1991; Canadian Petroleum Association, 1990; Jones and Leuterman, 1990; Wojtanowicz, 1993).

6.1 TREATMENT OF WATER

A number of methods are available to treat contaminated water to prepare it for reuse or disposal. The contaminants in water most commonly encountered in the petroleum industry can be grouped into two broad categories: hydrocarbons and solids. These contaminants can be either suspended or dissolved as discussed below.

6.1.1   Removal of Suspended Hydrocarbons

Suspensions of oil droplets in water (emulsions) can be difficult to separate because they can be stabilized by the interfacial energy between the oil droplets and the continuous water phase. A variety of methods are available to remove suspended droplets from water. These methods consist primarily of variations of gravitational separation, filtration, or biological degradation.

Gravity Separation

The first step in the removal of hydrocarbons from water is normally gravity separation. Through properly selected separator tanks with skimmers, most free oil and unstable oil emulsions can be separated from the water. Gravity separation is usually the simplest and most economical way to remove large quantities of free oil from water. However, more advanced methods are normally required to separate stable emulsions.

The first stage of gravity separation is to pass the water through large tanks to allow the phases to separate. These tanks are commonly called free water knockouts, wash tanks, settling tanks, or gun barrels. The effectiveness of these tanks depends on the droplet size and how long the water is in the tank (Arnold and Koszela, 1990; Arnold, 1983; Powers, 1990 and 1993). A schematic of a horizontal separator is shown in Figure 6-1.

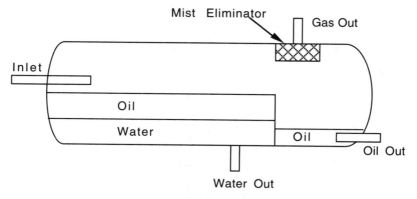

Figure 6-1. Schematic of a horizontal separator.

Plate separators can be used to improve the separation of oil and water. These separators consist of a series of closely spaced parallel plates that allow oil droplets to adhere to the plates, coalesce, and migrate along them. The closely spaced plates reduce the settling distance required to separate the oil droplets from the water. Plate separators are mechanically simple and require little maintenance. They are relatively large and are not effective for very small oil droplets. Plate separators can reduce oil concentrations to 2–25 mg/l, with an average of 15 mg/l (Simms et al., 1990), and can remove oil droplets down to about 20–30 micrometers in diameter (Van Den Broek and Plat, 1991). As summarized in Table 6-1, plate separators can have operational difficulties under some conditions.

Hydrocyclones can be used to further separate oil and water. A high-velocity stream is injected tangentially into the conically-shaped hydrocyclones, creating a vortex. The radial acceleration created in the hydrocyclone can be several orders of magnitude greater than that of gravity, and forces the more dense water to the outer edge of the hydrocyclone and the less dense oil to the center. The oil is then produced out of one end of the hydrocyclone and the water out of the

Table 6-1
Operational Problems with Oil Separation Equipment

Plate Separators	Hydrocyclones	Gas Flotation
Plugging of plates	Erosion	Unable to handle emulsions
Unable to handle emulsions	Corrosion	Level control problems
Platform motion	Sand buildup	Platform motion
Oil slugs		Oil slugs
Surge loads		Poor froth formation
		Interference by treatment chemicals
		Poor mechanical durability
		Scale/sludge buildup
		Operator/maintenance intensive

Source: Simms et al., 1990.

other. The effectiveness of hydrocyclones in separating oil and water depends on a large number of parameters, including oil droplet size and oil/water density difference, inlet water velocity, solution gas, solids, and system geometry (Flanigan et al., 1989; Jones, 1993; Meldrum, 1988; Smyth and Thew, 1990; Young et al., 1991b). Depending on the conditions, hydrocyclones can reduce oil concentrations to 10 ppm, but 30 ppm is a more common average (Simms et al., 1990). As summarized in Table 6-1, hydrocyclones can also have operational difficulties under some conditions. A schematic of a hydrocyclone is shown in Figure 6-2.

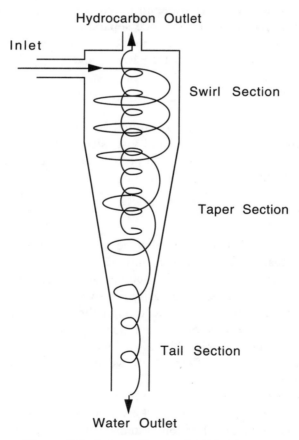

Figure 6-2. Schematic of a hydrocyclone.

Hydrocyclones for separating oil and water are limited to cases where the inlet pressure is sufficient to drive the flow (Flanigan et al., 1989). For low-pressure operations, the fluid may need to be pumped into the hydrocyclone. A progressive cavity pump with low shear has been found to be an effective way to increase the fluid pressure without shearing the oil into smaller drop sizes. The drop size is a critical parameter in the effectiveness of hydrocyclones in separating oil from water.

A related way to enhance gravity separation is through a decanting centrifuge. In this device, the produced water enters the spinning centrifuge, where the oil is separated from the water because of its lower density. Centrifuges differ from hydrocyclones in that the spinning is mechanically driven in a centrifuge, while it is induced by the inlet velocity of the water in a hydrocyclone. A centrifuge can also have internal plates to enhance separation, making it a spinning plate separator. Centrifuges can remove oil droplets down to about 2 micrometers in diameter (Van Den Broek and Plat, 1991).

Heater Treaters

Oil and water can also be separated by heating the mixture. The higher temperature lowers the fluid viscosity of the mixture and alters the interfacial tension between the phases, allowing the oil and water to separate faster.

Gas Flotation

Suspended oil droplets can also be removed from water by gas flotation. If gas bubbles are passed through an emulsion of oil-in-water, the oil droplets will attach to the bubbles and be carried to the top of the mixture where they can be easily removed. Air bubbles are normally pumped through the water, although the expansion of dissolved air is also used. Gas flotation is often aided by the addition of chemical coagulants. Carbon dioxide has also been used as the flotation gas (Burke et al., 1991). Gas flotation, however, can create a foam that is difficult to break.

Gas flotation systems can reduce oil concentrations to 15–100 mg/l, with a typical average of 40 mg/l (Simms et al., 1990). Like other separation methods, gas flotation systems can have operational difficulties, as summarized in Table 6-1.

Filtration

One way to remove oil droplets from water is to pass the water through water-wet filters or membranes. These filter media use capillary pressure to trap oil and prevent it from passing out of the filter. Advanced filtration processes include crossflow membranes such as microfiltration and ultrafiltration (Chen et al., 1991). These processes consist of a hydrophilic microfiltration membrane that passes water (and dissolved material), but not oil droplets. The shape of the filter is typically a small diameter capillary tube that the emulsion flows through. A schematic of a microfiltration capillary is shown in Figure 6-3. The emulsion leaving the tube without passing through the filter can be recycled through the filter a number of times to further concentrate the emulsion for other types of treatment or disposal. Microfiltration processes are usually ineffective for hydrocarbon removal, however, because the filters and membranes foul easily by oil and have short useful lifetimes.

Filtration Coalescence

Another type of filtration is to pass the water through oil-wet filters. The oil droplets attach to the filter matrix and coalesce into larger ones. When the filter medium has become saturated, larger oil drops will flow out of the filter, either by continued injection or by backwashing. These larger droplets can be more easily removed from the water by subsequent gravity separation. Sand, gravel, or glass fibers are common media used for this process.

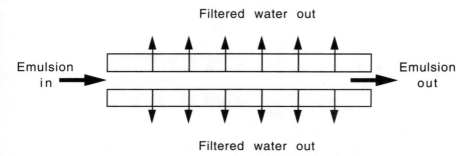

Figure 6-3. Schematic of a microfiltration capillary tube.

Chemical Coagulants

The removal of small, suspended oil droplets can be aided by adding chemicals that coagulate and flocculate the droplets (American Petroleum Institute, 1990). These chemicals typically overcome the electrostatic repulsion charges on the individual droplets, allowing them to coagulate into larger drops. These larger drops can then be more efficiently removed with gravity separation equipment. Common chemicals used include lime, alum, and polyelectrolytes. The use of dithiocarbamate has also been reported (Durham, 1993).

Electric Field Separation

Another way to separate oil from water is by applying an electric field (voltage) to the water to electrostatically remove the oil. These fields can be applied through either a direct or an alternating current. Oil droplets in an oil-in-water emulsion have a negative surface charge (zeta potential) that can be manipulated to facilitate their removal.

When a direct current is applied to the water containing such an emulsion, the oil will migrate toward the positive electrode. The migration velocity of the drops in many systems is on the order of 1 mm/min, which requires separators using very narrowly spaced parallel plates (Fang et al., 1991). This process, however, can only be used with saline water.

When an alternating current is applied, the droplets may flocculate if a metal hydroxide is present (Farrell, 1991). This process is known as *alternating current electrocoagulation.* In this process, a metal hydroxide is added to the water and an alternating current is used to overcome the electrostatic repulsion charges on the particles. When the electrostatic repulsion charges have been neutralized, the particles can flocculate and be more easily separated from the water by other methods. Iron and aluminum hydroxides have been successfully used.

Biological Processes

Biological processes rely on bacterial degradation of hydrocarbons. They have limited application in the removal of free hydrocarbons from most wastewater streams in the petroleum industry because they are too slow and are not appropriate for high oil concentrations. Large quantities of free oil can limit mass transfer of oxygen and nutrients to bacterial colonies that degrade the hydrocarbons. (American Petroleum Institute, 1986a). The application of biological processes to other waste streams will be discussed below.

6.1.2   Removal of Dissolved Hydrocarbons

In addition to suspended hydrocarbons, most produced water also contains varying amounts of dissolved hydrocarbons. A variety of methods are available to remove these dissolved hydrocarbons from the water.

Adsorption

An effective way to remove low levels of dissolved hydrocarbons is to adsorb it onto a solid medium. The most widely used medium is activated carbon. The pH and temperature of the system impacts the effectiveness of activated carbon on removing different hydrocarbon compounds. All free oil must be removed prior to the use of activated carbon to prevent the oil from clogging the carbon. In some cases, coal may also be used as an adsorption media. Natural and synthetic resins have also been developed that have proven effective in removing dissolved hydrocarbons from water.

Volatilization

Volatile organic carbon compounds (VOCs) can be removed from water by lowering the partial pressure of the compound in the vapor in contact with the water. When the partial pressure of the dissolved VOCs in the water exceeds that of its vapor pressure, the compounds will come out of solution and enter the vapor phase.

A variety of methods can be used to volatilize VOCs. Perhaps the most common is air stripping. In this process, air and water are passed through a containment vessel in countercurrent flow where VOCs evaporate into the air. The removal of VOCs can be enhanced by heating the air or by using steam, because higher temperatures increase their vapor pressure. Volatilization can also be enhanced by pulling a vacuum on the water, lowering the total system pressure.

One limitation to volatilization is that it transfers the VOCs from water to a vapor phase, yielding a contaminated vapor stream that must then be handled. If air is used, the oxygen will dissolve into the water, enhancing any biological degradation of dissolved hydrocarbons remaining in solution.

Biological Processes

Biological treatment can be used to remove low levels of dissolved hydrocarbons from wastewater streams. Biological treatment consists of mixing oxygen and nutrients with the water in a tank. The bacteria then degrade the organic compounds. This process is widely used in municipal water treatment plants, but may be too slow for oilfield applications. Because the high salinity of produced water inhibits biological growth, biological treatment will not be effective in most cases. Another limiting factor is the lack of dissolved oxygen for bacteria. Although oxygen could be added, it would significantly increase the corrosion rate of the equipment.

Precipitation

The solubility of many organic molecules decreases as the pH decreases. By lowering the pH, some organic materials can be precipitated. Precipitation, however, will not remove all dissolved hydrocarbons and will acidify the water.

Ultraviolet Irradiation

The use of ultraviolet radiation (including solar radiation) to break down hydrocarbons has also been studied (Green and Kumar, 1990). In this process, high-energy, short-wavelength photons are used to break the chemical bonds of dissolved hydrocarbons. When combined with heating to high temperatures, e.g., by solar collection panels, virtually complete destruction of hazardous hydrocarbon molecules in water has been observed. This method may have potential for treating some hazardous chemicals, but is probably too expensive for treating oilfield waters.

Oxidation

Dissolved hydrocarbons can also be destroyed through oxidation. Ozone, peroxide, chlorine, or permangenate have been tested. To be effective, however, oxidation normally must be conducted at high temperatures or with ultraviolet irradiation. Oxidation is not practical for most oilfield applications.

6.1.3 Removal of Suspended Solids

During many drilling and production activities, solids will be suspended in water that must be removed prior to water disposal. These solids include cuttings generated during drilling and sand and clay particles produced during oil production. Several methods are available for removing these suspended solids from the water.

Gravity Separation

The simplest way to separate the larger solid particles is to use gravitational settling. Fluids can be discharged into pits or tanks, where the solids settle to the bottom. Gravitational settling, however, is not effective for very small particles. The use of settling pits may also be limited by environmental regulations and the potential for future liability. Centrifuges can be used for enhanced gravitational separation.

Filtration

Another way to remove suspended solids is to filter the water. The water passes through the filter, while the solids are retained. The resulting filter cakes may be nonhazardous and could be disposed of like pit bottom sludge. Filtration has considerable promise for separating oil field wastes (Townley et al., 1989).

Coagulation

An effective way to enhance the separation of suspended particles is to coagulate (flocculate) the particles into larger agglomerations. The larger agglomerations can then be separated more easily by gravitational settling, centrifugation, or filtration.

One successful way to coagulate suspended solids is to add chemicals that overcome the electrostatic repulsive charges on the solids to allow them to flocculate. Chemicals that can be used include calcium chloride, ferric chloride, or aluminum potassium sulfate (Hinds et al.,

1986). A high molecular weight polyacrylamide polymer has been found to be effective to flocculate solids in water-based drilling muds, and a nonionic polyethylene oxide with a high molecular weight nonionic polyacrylamide polymer has been found to be effective for oil-based muds (Sharma and Smelley, 1991). Chemically enhanced centrifugation has been successfully used to remove solids from both drilling mud and produced water (Malachosky et al., 1991).

Suspended solids can also be flocculated with alternating current electrocoagulation (Farrell, 1991). In this process, a metal hydroxide is added to the water and an alternating current is used to overcome the electrostatic repulsion charges on the particles. Iron and aluminum hydroxides have been successfully used.

6.1.4   Removal of Dissolved Solids

Most wastewater also contains dissolved solids, particularly salt, hardness ions (calcium and magnesium), and heavy metals. A variety of methods are available to treat these waters. The methods vary considerably in cost and effectiveness.

Ion Exchange

Ion exchange (water softening) is an effective way to remove hardness ions from water. In most cases, the hardness ions (calcium and magnesium) are replaced with sodium ions. The removal of hardness ions is necessary for many processes because these ions readily precipitate and form a hard scale that can foul equipment.

There are two major ion exchange resins (substrates) that are commonly used: strong acid resins, using sulfonic acid, and weak acid resins, using carboxylic acid. Strong acid resins can be regenerated simply by flushing with a concentrated solution of sodium chloride. Weak acid resins, however, must be regenerated by flushing with a strong acid-like hydrochloric or sulfuric and then neutralizing with sodium hydroxide.

In some cases, the water can simply be passed through a bed of clay particles. The cation exchange capacity of most clays is very high, which allows them to trap and retain relatively high concentrations of dissolved metals. Activated alumina filtration is also an effective ion exchange media for metals like lead, mercury, and silver.

Precipitation

Many dissolved solids precipitate from water to form scale as the temperature, pressure, and/or chemistry changes. The most widely used system for precipitation is to add lime (CaOH) or sodium hydroxide (NaOH) to increase the pH of the water. At high pHs, dissolved solids, including heavy metals, tend to precipitate as a hydroxide sludge. Lime plus sodium carbonate can also be used to enhance the precipitation of calcium carbonate. The pH at which many metal hydroxides will precipitate is shown in Table 6-2.

Precipitation of some dissolved solids, particularly calcium and radium, can be enhanced by allowing the water to flow in channels open to the atmosphere (Caswell et al., 1992). Dissolved heavy metals can also be flocculated with organic materials to form colloids. These colloids can then be removed from the water as a suspended solid. Most forms of precipitation, however, leave residual levels of solids dissolved in solution. These residual levels may still exceed regulatory standards, and additional treatment of the water may be required.

Table 6-2
Precipitation of
Metal Hydroxides as
a Function of pH

Metal	pH
Al^{3+}	4.1
Cd^{2+}	6.7
Co^{2+}	6.9
Cr^{3+}	5.3
Cu^{2+}	5.3
Fe^{2+}	5.5
$Fe^{3}+$	2.0
Hg^{2+}	7.3
Mn^{2+}	8.5
Ni^{2+}	6.7
Pb^{2+}	6.0
Zn^{2+}	6.7

Source: Dean et al., 1972.
Copyright 1972, American Chemical
Society. Reprinted with permission.

Reverse Osmosis

The most common way to totally remove all dissolved solids from water is through filtration processes like reverse osmosis. These processes, however, are not intended to be used for wastewater treatment, but to provide potable water from nonpotable water. For example, reverse osmosis is commonly used to provide drinking water from seawater in desalinization plants. During reverse osmosis, saline water is pumped through a very small pore filter. The water molecules pass through the filter, but the larger dissolved solids molecules do not. Although the water supplied by a reverse osmosis plant is pure enough to be used for most purposes, the dissolved solids concentration in the waste stream that does not pass through the filter is higher than before and must still be disposed. Fouling is the most difficult problem to overcome when using reverse osmosis on oilfield brines. Pretreatment of the water prior to entering the reverse osmosis facility is required.

Because of its high cost, reverse osmosis is most commonly used to provide a supply of pure water in arid areas, rather than as a treatment method for wastewater. However, in areas where high-quality water is scarce, reverse osmosis can be used to treat produced water. (Tao et al., 1993).

Evaporation/Distillation

Another way to obtain potable water from water containing impurities is to evaporate and condense the water. Like reverse osmosis, this process is primarily used to provide a stream of pure water, not to treat a stream of wastewater. Like reverse osmosis, this process concentrates the wastes, which results in a smaller waste volume that ultimately must be disposed. This process is also very expensive.

Biological Processes

Although biological processes cannot destroy dissolved solids, they can alter their chemical form. For example, biological processes can alter the availability of heavy metals for uptake by plants, as well as the ability of metals to leach through the soil (Canarutto et al., 1991). Bacterial remediation has also been successfully used to remove sulfides from produced water (Sublette et al., 1993).

6.1.5 Neutralization

Many aqueous wastes in the petroleum industry are either acidic or alkaline. These wastes often must be treated to neutralize their reactivity before reuse or disposal. In many cases, the simplest treatment method is to mix these types of wastes for mutual neutralization. Because mixing may result in an exothermic reaction, it must be done with care to minimize any safety hazards.

6.2 TREATMENT OF SOLIDS

During drilling and production activities, a substantial volume of contaminated cuttings, soil, and produced solids are generated. The most common treatment method is to separate the solids from any contaminating water and/or hydrocarbons.

A variety of treatment methods are available to clean contaminated solids and are reviewed below. The effectiveness of different treatment methods depends on the solid type and size, as well as the initial contamination level and targeted final contamination level. Preprocessing techniques, including materials handling, can also impact the effectiveness of a treatment method. Preliminary tests of a particular method on a representative sample are recommended.

6.2.1 Removal of Water

A variety of methods are available to remove water from solids, including evaporation and filtration. One of the most common applications of dewatering technology is treating reserves pits containing drill cuttings and water-based drilling muds.

Evaporation

The simplest way to dewater solid wastes in arid climates is to put them in open pits or on concrete pads and allow the free water to evaporate. Evaporation is a common way to remove water from reserves pits following drilling, although changes in regulations may now require a more rapid dewatering than evaporation allows. Produced water can also be disposed of by evaporation, as long as the volumes are relatively low (Mutch, 1990).

In most cases, no special attempt has been made to limit leaching of metals or hydrocarbons from reserves pits or evaporation ponds. If leaching is a problem, the pit can be constructed with an impermeable liner and a leachate collection system with monitoring wells and enhanced evaporation features (sprinkler recirculation to increase the surface-to-volume ratio of the water). Lined pits are now required in some areas for oil-based or salty drilling mud systems. Any suspended or dissolved solids in the water will be concentrated as the water evaporates. If the pond has completely dried, these materials will be converted into a sludge, which may require further treatment before disposal.

Before dewatering and closure of reserves pits, the pit contents can segregate into layers. These layers can include a layer of free oil floating on a layer of water. The water normally contains a high concentration of dissolved solids. At the bottom is a layer of sludge that contains most of the settled solids. As the oil layer is weathered, a surface crust can also form. These top layers inhibit the evaporation of water, delaying the natural dewatering of the pits.

Percolation

In some arid areas where the water table is very deep, aqueous wastes can be placed in percolation ponds. These ponds have permeable sides and bottoms, allowing the water to percolate into the surrounding soil, leaving the solids at the bottom of the pond. The use of these ponds is highly restricted, however, because they allow dissolved solids in the water to spread into the surrounding soil.

Mechanical Methods

In many cases, evaporation is too slow to remove water from solid wastes. A number of mechanical methods are available to dewater solids. Preliminary separation of free liquids from the solids should be made with shale shakers, settling ponds, or hydrocyclones.

To further reduce the free water content of sludges, more advanced (and expensive) technologies can be used. These technologies include high-pressure filter presses, centrifuges, and vacuum filtering. Polymer conditioning of sludges can also be used to enhance dewatering. The low water content of the high-pressure filter presses can significantly lower disposal costs (Groves and Bartman, 1991a and 1991b; Mayer and Cregar, 1991; Steward, 1991).

The effectiveness of mechanical methods to dewater solids from a reserves pit and a production pit varies (Wojtanowicz et al., 1987). The dewatering of most oilfield wastes can be improved by preconditioning before mechanical separation with nonionic or low-charge anionic polymers with high molecular weights. Belt presses and centrifuges show similar performance, but belt presses are difficult to clean. Vacuum filtration and screw presses are not as effective because of their low volume reduction of the solid waste stream. Comparison of the effectiveness of several mechanical separations methods for reserves and production pits are provided in Tables 6-3 and 6-4, respectively. For these tables, initial solids contents of 10 wt% and 30.5 wt% were used, respectively.

Table 6-3
Effectiveness of Solids Separation Methods for Reserve Pits

	Belt Press	Centrifuge	Vacuum Filter
Volume Reduction (vol%)	70	71	45
Solids Recovery (wt%)	99.91	99.85	99.84
Cake Dryness (wt% solids)	45	44	23
Effluent Solids (mg/l)	150	180	130

Source: after Wojtanowicz et al., 1987.
Copyright SPE, with permission.

Table 6-4
Effectiveness of Solids Separation Methods for Production Pits

	Belt Press	Centrifuge	Screw Press
Volume Reduction (vol%)	50	38	26
Solids Recovery (wt%)	99.83	99.99	99.98
Cake Dryness (wt% solids)	55	59	45
Effluent Solids (mg/l)	300	30	95

Source: after Wojtanowicz et al., 1987.
Copyright SPE, with permission

6.2.2   Removal of Hydrocarbons

A variety of methods are available to remove hydrocarbons from solids, such as drill cuttings, contaminated soil, and produced sand. These methods include washing, adsorption, filtration, heating, solvent extraction, incineration, and biological degradation (U.S. Environmental Protection Agency, 1990). These methods are described below.

The effectiveness of these methods varies significantly. Pyrolysis can reduce most hydrocarbon concentrations on solids to nondetectable levels, while solvent extraction and distillation can reduce concentration to a few tens of mg/kg. Hydrocarbon concentrations following simple filtration can be in the hundreds of mg/kg range (American Petroleum Institute, 1987).

Washing

One of the least expensive ways to remove most of the hydrocarbons from solids is to wash them. The solids can be entrained in a fluidized bed of upward-flowing, high-velocity water. This stream agitates the solids and opens the pore system to release the oil. The efficiency of this process can be enhanced by adding a surfactant (soap) to the water to lower the interfacial tension holding the oil to the solids. Washing is more effective in sandy soils containing low amounts of clay.

A related process is to slurry the solids in a low-toxicity base oil. Although this process does not necessarily result in a lower hydrocarbon concentration in the solids, it can replace the original hydrocarbon, e.g., diesel, with a less toxic one.

If the volume of solids is small, they can be placed in an ultrasound bath for cleaning. The high-frequency acoustic pulses in the bath help release the hydrocarbons from the solids. Ultrasound baths work well for laboratory scale operations, but are not appropriate for oilfield-scale applications.

Adsorption

Another relatively low-cost method of removing some of the hydrocarbons contaminating solids is to mix the soil with a material that is strongly oil-wet, like coal or activated carbon (Ignasiak et al., 1990).

A suspension of contaminated soil and the carbon can be tumbled in water at elevated temperatures to allow the oil to be absorbed by the carbon. The oily carbon is then separated from the water and clean sand by flotation. The oily carbon can then be burned in conventional coal-fired power plants or buried in an approved facility.

Heating

Heating cuttings contaminated with hydrocarbons can help separate the hydrocarbons from the solids, particularly when being washed in water (Henriquez, 1990). This procedure is similar to using heat to break emulsions and separate hydrocarbons and water.

Heating can also be used for hydrocarbon sludges (Hahn, 1993). In this process, the sludges are heated above the boiling point of water and allowed to flash to vapor. This separates the water and light hydrocarbons from the heavier hydrocarbons. The high temperature also lowers the viscosity of the heavy hydrocarbons, facilitating their separation as a slurry.

Distillation/Pyrolysis

A more expensive method for removing light- and intermediate-weight hydrocarbon compounds is to distill them from the solids in a retort furnace. The solid/hydrocarbon mixture is heated to vaporize the light and medium molecular weight hydrocarbons and water. The gases are removed from the high-temperature chamber by either a nitrogen or steam sweep. After the vapors are subsequently cooled and condensed, the oil is separated from the water. The oil can be reused and the solids and water sent to an appropriate disposal facility. To maximize the separation of liquids and solids, the heating can be done in a rotating drum with hammers to crush the solids while rotating. Several commercial thermal distillation processes are available (Ruddy et al., 1990).

Distillation systems, however, have several significant operations limitations. Hydrocarbon vapors at high temperatures are a fire hazard, corrosion problems increase significantly at high temperatures, and air pollutants are emitted. The chemical structure of some hydrocarbons is altered at high temperatures, making their reuse in some applications, like drilling muds, impossible. If heavy hydrocarbon components are present, they will not be distilled and will form a heavy residual tar on the solids. For example, distillation may remove only about 65% of the heavy polynuclear aromatics. Another limitation to distillation is the high energy costs of heating the materials to a sufficiently high temperature. An operating temperature of about 800°F may be required for effective distillation of heavy ends (Young et al., 1991a). An operating temperature of 473°F, however, has proven to be effective in lowering the hydrocarbon level of cuttings to 10 g/kg (Van Elsen and Smits, 1991).

If the distillation temperatures are high enough, the hydrocarbon molecules will be broken by pyrolysis, forming coke. This would solidify the remaining hydrocarbons, preventing their migration upon disposal of the waste.

Incineration

Another way to remove hydrocarbons from solids is to burn the mixture in an incinerator. Incinerators are specially designed burners that can burn the relatively small volume of combustible materials found in oily solids. Following combustion, the resulting ash, including any salts and heavy metals, is solidified to prevent leaching of any hazardous residue. Incineration typically removes over 99% of the hydrocarbons in the soil.

A significant limitation to incinerators is that they emit air pollutants, particularly metal compounds like barium, cadmium, chromium, copper, lead, mercury, nickel, vanadium, and zinc. Incineration destroys hydrocarbon wastes, but merely changes the chemical form of heavy metal wastes. Because of the air pollutants emitted, all incinerators require permits. Another limitation to incineration is that a secondary fuel is required because the heat content of the hydrocarbons in many petroleum solid wastes is insufficient for combustion, particularly when a high volume of noncombustible material is present, e.g., the solids. The need for secondary fuel increases the cost of operations. Although incineration is expensive, it has low future liability (Goodwin and Turner, 1990).

Solvent Extraction

Solvent processes can also be used to separate hydrocarbons from solids. In these processes, a solvent with a low boiling point is mixed with the oily solids to wash the oil from the solids and dilute what remains trapped. The solvent is then separated from the hydrocarbons and solids by low-temperature distillation and reused. Solvent extraction is routinely used in the petroleum industry for extracting fluids from cores during core analysis. Like distillation, solvent extraction is expensive. Solvent extraction is more effective in sandy soils containing little clay. Several commercial solvent extraction processes are available (Ruddy et al., 1990).

A more exotic solvent extraction process uses critical or supercritical fluids. In this process, the cuttings are placed in a pressure chamber with a fluid near its critical point. Commonly used fluids include carbon dioxide, propane, ethane, and butane. The pressure is increased until the fluid passes above its critical point and becomes a liquid. The liquid is then used as a solvent to wash the oil from the solids. After the liquid mixture is separated from the solids, the pressure is lowered. With the lower pressure, the supercritical fluid reverts to a gaseous state, leaving the extracted hydrocarbons behind. The gas is then recycled. The process is expensive, but eliminates many of the problems associated with high-temperature thermal processes.

Biological Processes

Most hydrocarbons encountered in the upstream petroleum industry can be biologically converted to carbon dioxide and water by microbes like bacteria and fungi. During biological degradation, the hydrocarbons are eaten as food by the bacteria. This biological degradation can be enhanced by providing the optimum conditions for microbe growth. The deliberate enhancement of biological degradation is called *bioremediation.*

The effectiveness and speed of bioremediation in degrading hydrocarbons depends on a variety of environmental conditions, including temperature, salinity, pH, hydrocarbon type, heavy metal concentration, soil texture, moisture content, and hydrocarbon concentration. Because of this, the chemical composition of the hydrocarbon, the type and level of background microorganisms, and the nutrient level at the site must be determined and the environmental conditions controlled for optimum degradation (American Petroleum Institute, 1986b; Hildebrandt and Wilson, 1991).

Naturally occurring bacteria can effectively degrade naturally occurring hydrocarbons, such as crude oil. In most cases, the appropriate bacteria are already present in the environment and their populations can be increased just by adding nutrients. In some cases, naturally occurring bacteria have been artificially cultured and then released in greater numbers to accelerate biodegradation of the hydrocarbons, but the effectiveness of this augmentation is uncertain. Genetically engineered bacteria may be necessary to degrade some refined hydrocarbons, such as chlorinated solvents.

The most significant limitation for many bioremediation applications is a lack of nutrients for bacterial growth. These nutrients, e.g., nitrogen, phosphorus, and some trace elements, can be added by way of fertilizer. The amount and composition of fertilizer needed for optimum degradation depends on what hydrocarbon is being degraded and the bacteria being enhanced.

Oxygen is also needed for bioremediation to convert the hydrocarbons to carbon dioxide and water. Anaerobic biological degradation (without oxygen) also occurs, but is much slower and less efficient than aerobic degradation. Oxygen is normally provided by ensuring that the pore system within the solids is sufficiently open for air to flow through it. One way to enhance the pore system is by adding inert bulking agents like wood chips, bark, sawdust, tires, and shredded vegetation to increase the mixture porosity. The use of inert bulking agents is called *composting bioremediation.*

In most cases, water is also needed because it is the medium in which the bacteria live. Bacterial growth normally occurs at water/hydrocarbon interfaces. For optimum degradation, the water content of the solids must be balanced. If not enough water is present, bacterial growth will be inhibited. If too much water is present, the access of oxygen and nutrients to the bacteria will be limited, again inhibiting bacterial growth.

In some cases, surfactants have been added to the nutrient mixture to solubilize and emulsify low-solubility hydrocarbons, including heavy aromatics and PAHs. Surfactants can also mobilize sorbed microbial cells and contaminants from the soil surface to provide greater access to microbial attack.

The degradation rate of hydrocarbons depends on the structure of the hydrocarbon molecule and the type of bacteria involved. Paraffins are the most susceptible to microbial attack, followed by isoparaffins and aromatics. The polycyclic aromatic hydrocarbons (PAHs) are the most difficult to biodegrade. The speed of bioremediation is often measured in terms of half-lives of the hydrocarbon, i.e., the time for half of the hydrocarbon to be biologically degraded. Typical degradation half-lives range from a few days for low-molecular-weight compounds to a number of years for complex compounds (American Petroleum Institute, 1984). When the oil has been degraded and hydrocarbon levels have been reduced, the bacterial populations return to their initial level.

Specific bioremediation half-lives have been reported as over 48 weeks for bunker C fuel (Song et al., 1990), 37–57 weeks for crude oil sludge (Loehr et al., 1992), less than 30 days for some normal alkanes and aromatics (Loehr et al., 1992), five weeks for a Saudi Arabian crude oil (Whiteside, 1993), eight weeks for crude oil under optimum conditions (McMillen et al., 1993), and more than two years for crude oil under nonoptimized conditions (McMillen et al., 1993. Bioremediation with composting has also been successfully applied with remediation times of five weeks for sludges and diesel-contaminated soils (Martinson et al., 1993).

There are a number of difficulties, however, with applying bioremediation at hydrocarbon-contaminated sites. The presence of dissolved solids, such as heavy metals or salt in reserves pits, will inhibit bacterial growth. Bioremediation projects can emit significant levels of air pollutants. For example, the emission rates of benzene and naphthalene may be high enough to require that respirators be worn by workers (Myers and Barnhart, 1990). Such air emissions are expected to limit the availability of permits for bioremediation projects in the future. Finally, some bioremediation facilities have required large amounts of water, which can be a problem in arid areas.

Filtration

If the hydrocarbon content of the solids is high, some of the free hydrocarbons can be separated from the solids by mechanical filtration. Filtration, however, is not effective for reducing hydrocarbon concentrations to low levels.

6.2.3   Solidification

One way to treat contaminated solids is to solidify the mixture so that the contaminants become part of the solid. Solidification reduces pollutant mobility and improves handling characteristics. Two types of solidification have been used: adding materials to absorb free liquids and adding materials to chemically bind and encapsulate the contaminants. Most off-site disposal sites use solidification to treat the wastes for final disposal by burial (Jones, 1990; Roberts and Johnson, 1990).

Absorbants are typically used to dewater reserves pits in areas where the evaporation rate is low. Materials that have been added to the pits to absorb free water include straw, dirt, fly ash, clays, kiln dust, fly ash, and polymers.

The best solidification methods, however, are those that chemically bind the contaminants. These methods are based primarily on portland cement, calcium silicate, or alumino-silicate reactions (Carter, 1989; Nahm et al., 1993). These materials, unlike fly ash or kiln dust, can reduce the leachability of toxic heavy metals, asbestos, oils, and salts. The mobility of metals from such solidification can be reduced by 80-90%, while that of organics can be reduced by 60-99% (U.S. Environmental Protection Agency, 1990).

Vitrification by heating the solids to a high enough temperature to melt silica has also been proposed (Buelt and Farnsworth, 1991), but is likely to be too expensive for applications in the petroleum industry.

6.3   TREATMENT OF AIR EMISSIONS

During drilling and production activities, a substantial volume of air pollutants can be generated and emitted. These pollutants include hydrocarbons, sulfur oxides, nitrogen oxides, and particulates. A variety of treatment methods are available, but their effectiveness varies considerably with the pollutant being treated.

6.3.1   Hydrocarbons

The primary source of hydrocarbon emissions is from the exhaust of internal combustion engines. Unfortunately, there is little that can be done to treat these emissions other than to operate the engines within their design specifications.

The vapor space in production tanks can collect volatile hydrocarbon vapors. These vapors can be collected and treated with vapor recovery systems (Webb, 1993). Casing gas from thermal enhanced oil recovery operations may also contain high levels of hydrocarbon vapors. These casing gases can be collected in a separate gathering system and treated by adsorption (Peavy and Braun, 1991).

Another source of hydrocarbon emissions are the fugitive emissions arising from leaking valves and fittings. Because these emissions are generally too spread out to be collected, their release must be prevented by replacing and repairing the leaking equipment.

Emissions from remediation projects of hydrocarbon-contaminated sites can contain volatile hydrocarbons. These hydrocarbons can be collected by passing the emissions through a bed of activated carbon or adsorptive polymer. Alternatively, the vapors can be bubbled through water, where the hydrocarbons become dissolved. Although the dissolution process can be effective in lowering hydrocarbon air emissions, the subsequently contaminated water must then be treated and disposed. For some projects, catalytic oxidation may be used as a low-temperature alternative to incineration of volatile hydrocarbons.

6.3.2   Sulfur Oxides

Sulfur oxides are generated from the combustion of fuels containing sulfur. Although these emissions can be treated to remove the sulfur, the emission of sulfur can also be reduced or eliminated by the use of low-sulfur fuel. A variety of scrubber systems are available to remove sulfur from air emissions (Goodley, 1979).

6.3.3   Nitrogen Oxides

Nitrogen oxides are generated from high-temperature combustion and from the combustion of fuels containing nitrogen (crude oil). Unfortunately, these emissions are difficult to treat and may require specially designed equipment.

Equipment to minimize the emission of nitrogen oxide in combustion gases includes low NO_x burners, flue gas recirculators, selective catalytic reduction devices, and selective noncatalytic systems. The amount of nitrogen oxides emitted can also be lowered by reducing the amount of oxygen in the combustion process. Unfortunately, lowering oxygen in the combustion process increases the amount of partially burned hydrocarbons created.

The impact of nitrogen oxides from fixed installations, such as natural gas compressor stations, can be minimized by the stack height, location, and orientation with respect to other structures (Ramsey and Roger, 1991).

6.3.4   Particulates

Many combustion operations emit partially burned hydrocarbon particulates from incomplete combustion. These particulates, such as soot, can be removed by passing the flue gas through a scrubber, where the particulates become entrained in the water.

REFERENCES

American Petroleum Institute, "The Land Treatability of Appendix VIII Constituents Present in Petroleum Industry Wastes,"API Publication 4379, Washington, D.C., 1984.

American Petroleum Institute, "Bacterial Activity in Ground Waters Containing Petroleum Products,"API Publication 4211, Washington, D.C., 1986a.

American Petroleum Institute, "Enhancing the Microbial Degradation of Underground Gasoline by Increasing Available Oxygen,"API, Publication 4428, Washington, D.C., 1986b.

American Petroleum Institute, "Evaluation of Treatment Technologies for Listed Petroleum Refinery Wastes,"API, Publication 4465, Washington, D.C., Dec. 1987.

American Petroleum Institute, "Monographs on Refinery Environmental Control-Management of Water Discharges,"API, Publication 420, Washington, D.C., Aug. 1990.

Arnold, K. E., "Design Concepts for Offshore Produced-Water Treating and Disposal Systems," *J. Pet. Tech.,* Feb. 1983, pp. 276–283.

Arnold, K. E. and Koszela, P. J., "Droplet-Settling vs. Retention-Time Theories for Sizing Oil/Water Separator," *SPE Production Engineering,* Feb. 1990, pp. 59–64.

Buelt, J. L. and Farnsworth, R. K., "In Situ Vitrification of Soils Containing Various Metals," *Nuclear Technology,* Vol. 96, Nov. 1991, pp. 178–184.

Burke, N. E., Curtice, S., Little, C. T., and Seibert, A. F., "Removal of Hydrocarbons From Oil Field Brines by Flocculation with Carbon Dioxide," paper SPE 21046 presented at the Society of Petroleum Engineers International Symposium on Oilfield Chemistry, Anaheim, CA, Feb. 20–22, 1991.

Canadian Petroleum Association, "Production Waste Management Handbook for the Alberta Petroleum Industry," Dec. 1990.

Canarutto, S., Petruzzelli, G., Lubrano, L., and Vigna Guidi, B., "How Composting Affects Heavy Metal Content," *Biocycle,* June 1991, pp. 48–50.

Carter, E. E., "Chemical Solidification of Hazardous Sludges and Other Wastes," Proceedings of Petro-Safe '89, Houston, TX, Oct. 3–5, 1989.

Caswell, P. C., Gelb, D., Marinello, S. A., Emerick, J. C., and Cohen, R. R. H., "Testing of Man-Made Overland-Flow and Wetlands Systems for the Treatment of Discharged Waters from Oil and Gas Production Operations in Wyoming," Proceedings of Petro-Safe '92, Houston, TX, 1992.

Chen, A. S. C., Flynn, J. T., Cook, R. G., and Casaday, A. L., "Removal of Oil, Grease, and Suspended Solids from Produced Water with Ceramic Crossflow Microfiltration," *SPE Production Engineering,* May 1991, pp. 131–136.

Dean, J. G., Bosqui, F. L., and Lanouette, K. H., "Removing Heavy Metals from Waste Water," *Environmental Science & Technology,* Vol. 6, No. 6, June 1972, pp. 518–522.

Durhan, D. K., "Advances in Water Clarifier Chemistry for Treatment of Produced Water on Gulf of Mexico and North Sea Offshore Production Facilities," paper SPE 26008 presented at the Society of Petroleum Engineers/Environmental Protection Agency's Exploration and Production Environmental Conference, San Antonio, TX, March 7–10, 1993.

Fang, C. S., Tong, N. A. M., and Lin, J. H., "Removal of Emulsified Crude Oil From Produced Water by Electrophoresis" paper SPE 21047 presented at the Society of Petroleum Engineers International Symposium on Oilfield Chemistry, Anaheim, CA, Feb. 20–22, 1991.

Farrell, C. W., "Oilfield Process Stream Treatment by Means of Alternating Current Electrocoagulation," *Advances in Filtration and Separation Technology, Vol. 3, Pollution Control Technology for Oil and Gas Drilling and Production Operations,* American Filtration Society. Houston: Gulf Publishing Co., 1991, pp. 186–207.

Flanigan, D. A., Skilbeck, F., Stolhand, J. E., and Shimoda, E., "Use of Low-Shear Pumps in Conjunction with Hydrocyclones for Improved Performance in the Cleanup of Low-Pressure Produced Water," paper SPE 19743 presented at the Society of Petroleum Engineers 64th Annual Technical Conference and Exhibition, San Antonio, TX, Oct. 8–11, 1989.

Freeman, H. M., editor, *Standard Handbook of Hazardous Waste Treatment and Disposal.* New York: McGraw-Hill Book Company, 1989.

Goodley, A., "Air Quality Impact of Thermally-Enhanced Heavy Oil Recovery in California, USA," proceedings of The Future of Heavy Crude Oils and Tar Sands, First International UNITAR Conference, Edmonton, Alberta, June 4–12, 1979.

Goodwin, S. and Turner, L. R., "The AOSTRA Taciuk Process—The Flexible Alternative for Oily Waste Treatment," paper No. 24 of the proceedings of Oil Sands 2000, AOSTRA, Edmonton, Alberta, March 26–28, 1990.

Green, K. M. and Kumar, D., "The Potential for Solar Detoxification of Hazardous Wastes in the Petroleum Industry," Proceedings of the

U.S. Environmental Protection Agency's First International Symposium on Oil and Gas Exploration and Production Waste Management Practices, Sept. 10–13, New Orleans, LA, 1990, pp. 771–781.

Groves, R. and Bartman, G. H., "EXXFLOW/EXXPRESS Microfiltration Membrane Technology for Treatment of Oil Produced Water," *Advances in Filtration and Separation Technology, Vol. 3, Pollution Control Technology for Oil and Gas Drilling and Production Operations,* American Filtration Society. Houston: Gulf Publishing Co., 1991a, pp. 141–148.

Groves, R. and Bartman, G. H., "Solids Dewatering Using EXXPRESS," *Advances in Filtration and Separation Technology, Vol. 3, Pollution Control Technology for Oil and Gas Drilling and Production Operations,* American Filtration Society. Houston: Gulf Publishing Co., 1991b, pp. 424–427.

Hahn, W. J., "High-Temperature Reprocessing of Petroleum Oily Sludges," paper SPE 25931 presented at the Society of Petroleum Engineers/Environmental Protection Agency's Exploration and Production Environmental Conference, San Antonio, TX, March 7–10, 1993.

Henriquez, L. R., "The Development of an OBM Cutting Cleaner in the Netherlands," Proceedings of the U.S. Environmental Protection Agency's First International Symposium on Oil and Gas Exploration and Production Waste Management Practices, New Orleans, LA, Sept. 10–13, 1990, pp. 243–254.

Hildebrandt, W. W. and Wilson, S. B., "On-Site Bioremediation Systems Reduce Crude Oil Contamination," *J. Pet. Tech.,* Jan. 1991, pp. 18–22.

Hinds, A. A., Donovan, D. M., Lowell, J. L., and Liao, A., "Treatment Reclamation and Disposal Options for Drilling Muds and Cuttings," paper IADC/SPE 14798 presented at the Society of Petroleum Engineers IADC/SPE Drilling Conference, Dallas, TX, Feb. 10–12, 1986.

Ignasiak, T., Carson, D., Szymocha, K., Pawlak, W., and Ignasiak, B., "Clean-Up of Oil Contaminated Solids," Proceedings of the U.S. Environmental Protection Agency's First International Symposium on Oil and Gas Exploration and Production Waste Management Practices, New Orleans, LA, Sept. 10–13, 1990, pp. 159–168.

Jones, F. V., "State Regulatory Programs for Drilling Fluids Reserve Pit Closure: An Overview," Proceedings of the U.S. Environmental Protection Agency's First International Symposium on Oil and Gas Exploration and Production Waste Management Practices, New Orleans, LA, Sept. 10–13, 1990, pp. 911–924.

Jones, P. S., "A Field Comparison of Static and Dynamic Hydrocyclones," *SPE Production and Facilities,* May 1993, pp. 84–90.

Jones, F. V. and Leuterman, A. J. J., "State Regulatory Programs for Drilling Fluids Reserve Pit Closure: An Overview," Proceedings of the U.S. Environmental Protection Agency's First International Symposium on Oil and

Gas Exploration and Production Waste Management Practices, Sept. 10–13, New Orleans, LA, 1990, pp. 911–924.

Loehr, R. C., Martin, J. H., and Neuhauser, E. F., "Land Treatment of an Aged Oily Sludge-Organic Loss and Change in Soil Characteristics," *Water Research,* Vol. 26, No. 6, 1992, pp. 805–815.

Malachosky, E., Sanders, R., and McAuley, L., "Impact of Dewatering Technology on the Cost of Drilling-Waste Disposal," *J. Pet. Tech.,* June 1991, pp. 730–736.

Martinson, M. M., Malter, P. L., McMillin, T. G., Fyock, L., and Wade, M., "Composting Bioremediation for Exploration and Production Wastes," paper SPE 26395 presented at the Society of Petroleum Engineers 68th Annual Technical Conference and Exhibition, Houston, TX, Oct. 3–6, 1993.

Mayer, E. and Cregar, D. E., "Improved Dewatering Techniques for Producing Burnable Wastes," *Advances in Filtration and Separation Technology, Vol. 3, Pollution Control Technology for Oil and Gas Drilling and Production Operations,* American Filtration Society. Houston: Gulf Publishing Co., 1991, pp. 358–376.

McMillen, S. J., Kerr, J. M., and Gray, N. R., "Microcosm Studies of Factors That Influence Bioremediation of Crude Oil in Soils," paper SPE 25981 presented at the Society of Petroleum Engineers/Environmental Protection Agency's Exploration and Production Environmental Conference, San Antonio, TX, March 7–10, 1993.

Meldrum, N., "Hydrocyclones: A Solution to Produced-Water Treatment," *SPE Production Engineering,* 1988, pp. 669–676.

Mutch, G. R. P., "Environmental Protection Planning for Produced Brine Disposal in Southwestern Saskatchewan Natural Gas Fields," Proceedings of the U.S. Environmental Protection Agency's First International Symposium on Oil and Gas Exploration and Production Waste Management Practices, Sept. 10–13, 1990, New Orleans, LA, pp. 375–386.

Myers, J. M. and Barnhart, M. J., "Pilot Bioremediation of Petroleum Contaminated Soil," Proceedings of the U.S. Environmental Protection Agency's First International Symposium on Oil and Gas Exploration and Production Waste Management Practices, New Orleans, LA, Sept. 10–13, 1990, pp. 745–756.

Nahm, J. J., Javanmardi, K., Cowan, K. M., and Hale, A. H., "Slag Mix Mud Conversion Cementing Technology: Reduction of Mud Disposal Volumes and Management of Rig-Site Drilling Wastes," paper SPE 25988 presented at the Society of Petroleum Engineers/Environmental Protection Agency's Exploration and Production Environmental Conference, San Antonio, TX, March 7–10, 1993.

Peavy, M. A. and Braun, J. E., "Control of Waste Gas From a Thermal EOR Operation," *J. Pet. Tech.,* June 1991, pp. 656–661.

Powers, M. L., "Analysis of Gravity Separation in Freewater Knockouts," *SPE Production Engineering,* 1990, pp. 52–58.

Powers, M. L., "New Perspectives on Oil and Gas Separator Performance," *SPE Production and Facilities,* May 1993, pp. 77–83.

Ramsey, S. H. and Roger, P. L., "Ambient Air Quality Impacts of Natural Gas Compressor Stations," paper SPE 21723 presented at the Society of Petroleum Engineers Production Operations Symposium, Oklahoma City, OK, April 7–9, 1991.

Roberts, L. and Johnson, G., "A Study of the Leachate Characteristics of Salt Contaminated Drilling Wastes Treated with a Chemical Fixation Solidification Process," Proceedings of the U.S. Environmental Protection Agency's First International Symposium on Oil and Gas Exploration and Production Waste Management Practices, New Orleans, LA, Sept. 10–13, 1990, pp. 933–944.

Ruddy, D., Ruggerio, D. D., and Kohlmann, H. J., "An Overview of Treatment Technologies for Reduction of Hydrocarbon Levels in Drill Cuttings Wastes," Proceedings of the U.S. Environmental Protection Agency's First International Symposium on Oil and Gas Exploration and Production Waste Management Practices, New Orleans, LA, Sept. 10–13, 1990, pp. 717–730.

Schmidt, E. and Jaeger, S., "PRS Treatment and Reuse of Oilfield Wastewaters," Proceedings of the U.S. Environmental Protection Agency's First International Symposium on Oil and Gas Exploration and Production Waste Management Practices, New Orleans, LA, Sept. 10–13, 1990, pp. 795–808.

Sharma, S. K. and Smelley, A. G., "Use of Flocculants in Dewatering of Drilling Muds," *Advances in Filtration and Separation Technology, Vol. 3, Pollution Control Technology for Oil and Gas Drilling and Production Operations,* American Filtration Society. Houston: Gulf Publishing Co., 1991, pp. 43–51.

Simms, K., Kok, S., and Zaidi, A., "Alternative Processes for the Removal of Oil from Oilfield Brines," Proceedings of the U.S. Environmental Protection Agency's First International Symposium on Oil and Gas Exploration and Production Waste Management Practices, New Orleans, LA, Sept. 10–13, 1990, pp. 17–30.

Smyth, I. C. and Thew, M. T., "The Use of Hydrocyclones in the Treatment of Oil Contaminated Water Systems," Proceedings of the U.S. Environmental Protection Agency's First International Symposium on Oil and Gas Exploration and Production Waste Management Practices, New Orleans, LA, Sept. 10–13, 1990, pp. 1001–1012.

Song, H. G., Wang, X., and Bartha, R., "Bioremediation Potential of Terrestrial Fuel Spills," *Applied and Environmental Microbiology,* Vol. 56, No. 3, March 1990, pp. 652–656.

Steward, C. R., "Minimizing Waste, Utilizing- High Pressure Membrane Filter Presses," *Advances in Filtration and Separation Technology, Vol. 3, Pollution Control Technology for Oil and Gas Drilling and Production Operations,* American Filtration Society. Houston: Gulf Publishing Co., 1991, pp. 396–405.

Sublett, K. L., Morse, D. E., and Raterman, K. T., "A Field Demonstration of Sour Produced Water Remediation by Microbial Treatment," paper SPE 26396 presented at the Society of Petroleum Engineers 68th Annual Technical Conference and Exhibition, Houston, TX, Oct. 3–6, 1993.

Tao, F. T., Curtice, S., Hobbs, R. D., Sides, J. L., Wieser, J. D., Dyke, C. A., Tuohey, D., and Pilger, P. F., "Conversion of Oilfield Produced Water into an Irrigation/Drinking Quality Water," paper SPE 26003 presented at the Society of Petroleum Engineers/Environmental Protection Agency's Exploration and Production Environmental Conference, San Antonio, TX, March 7–10, 1993.

Tchobanoglous, G. and Burton, F. L., *Wastewater Engineering: Treatment, Disposal, and Reuse.* New York: McGraw Hill, Inc., 1991.

Townley, D., Bergman, R. J., and Goldman, W. A., "Reduction of Oilfield Production Wastes Using Dry Cake Filters," paper SPE 19742 presented at the Society of Petroleum Engineers 64th Annual Technical Conference and Exhibition, San Antonio, TX, Oct. 8–11, 1989.

U.S. Environmental Protection Agency, "Summary of Treatment Technology Effectiveness for Contaminated Soil," PB92-963351, June 1990.

Van Den Broek, W. M. G. T. and Plat, R., "Characteristics and Possibilities of Some Techniques for De-Oiling of Production Water," paper SPE 23315 presented at the Society of Petroleum Engineers First International Conference on Health, Safety, and Environment, The Hague, Netherlands, Nov. 10–14, 1991.

Van Elsen, R. P. H. and Smits, M., "Cutting Cleaner: A Long-Term Field Test," paper SPE 23360 presented at the Society of Petroleum Engineers First International Conference on Health, Safety, and Environment, The Hague, Netherlands, Nov. 10–14, 1991.

Webb, W. G., "Vapor Jet System: An Alternative Vapor Recovery Method," paper SPE 25942 presented at the Society of Petroleum Engineers/Environmental Protection Agency's Exploration and Production Environmental Conference, San Antonio, TX, March 7–10, 1993.

Whiteside, S. E., "Biodegradation Studies of Saudi Arabian Crude Oil," paper SPE 26399 presented at the Society of Petroleum Engineers 68th Annual Technical Conference and Exhibition, Houston, TX, Oct. 3–6, 1993.

Wojtanowicz, A. K., "Oilfield Environmental Control Technology: A Synopsis," *J. Pet. Tech.,* Feb. 1993, pp. 166–172.

Wojtanowicz, A. K., Field, S. D., and Osterman, M. C., "Comparison Study of Solid/Liquid Separation Techniques for Oilfield Pit Closures," *J. Pet. Tech.,* July 1987, pp. 845–856.

Young, G. A., Growcock, F. B., Talbot, K. J., Lees, J., and Worrell, B., "Elements of Thermally Treated Oil-Base Mud Cuttings," paper SPE/IADC 21939 presented at the Society of Petroleum Engineers 1991 Drilling Conference, Amsterdam, The Netherlands, March 11–14, 1991a.

Young, G. A., Wakley, W. D., Taggart, D. L., Andrews, S. L., and Worrell, J. R., "Oil-Water Separation Using Hydrocyclones, An Experimental Search for Optimum Dimensions," *Advances in Filtration and Separation Technology, Vol. 3, Pollution Control Technology for Oil and Gas Drilling and Production Operations,* American Filtration Society. Houston: Gulf Publishing Co., 1991b, pp. 102–111.

CHAPTER 7

Waste Disposal Methods

The upstream petroleum industry generates a significant volume of wastes, primarily produced water and drill cuttings. No matter how effective a waste management plan or waste treatment program may be, wastes will remain that must be disposed of. In some cases, the final disposal can be on-site, while in other cases, the wastes must be shipped for disposal off-site.

Ultimately, petroleum industry wastes can be disposed of above or below the surface of either land or water. The suitability of these disposal locations varies with the wastes being disposed.

7.1 SURFACE DISPOSAL

The easiest and least expensive method of waste disposal is to discharge the wastes onto the ground or into surface waterways. Although this has historically been a common disposal method for many wastes, its use and misuse has been a major factor in the increase in environmental regulations governing the petroleum industry. Nevertheless, various forms of surface disposal are still appropriate for many treated wastes.

7.1.1 Disposal of Water

Wastewater can be discharged directly into local streams, rivers, or the ocean as long as its quality meets regulatory standards, i.e., its concentration of suspended and dissolved solids, chemicals, and hydrocarbons is sufficiently low. Surface discharge is regulated in most areas, however, and permits for such discharge are required.

When wastewater is discharged offshore, the water is typically treated to remove only the hydrocarbons. Although the dissolved solids (salt) concentrations of most produced waters are high enough to be toxic to even marine life, the rapid mixing and dilution of the discharged water makes the resulting environmental impact negligible.

For near-shore discharges in shallow water, there is less opportunity for mixing and dilution of the discharged water, and a toxic plume can exist for some distance away from the discharge point. Such toxic plumes are of particular concern when discharging a dense, high-saline, oxygen-deficient brine because it can be trapped in subsurface topographic low areas. Because this trapped brine can significantly impact the local marine life, permits to discharge high-salinity brines near the shore may be difficult to obtain, even if the hydrocarbon content is low.

When wastewater is discharged into onshore freshwater locations, both the hydrocarbon and dissolved solids concentrations must be low. Because of the high cost of removing dissolved solids, surface discharge of wastewater is generally possible only if the initial dissolved solids concentration of the water is low. Surface discharge into dry stream beds is a common way to dispose of treated water in arid areas like Wyoming.

Surface discharge into percolation ponds is also used in some areas. In percolation ponds, the water is allowed to percolate into the under-saturated (vadose) zone, where it eventually evaporates back into the atmosphere. Because of the lack of control over where the water goes, this disposal method is being phased out. Discharge into evaporation ponds is also an option in many arid areas, particularly if a liner is used to prevent leaching of dissolved solids.

7.1.2   Disposal of Solids

Waste solids can be discharged directly onto the ground or into the ocean as long as their quality meets regulatory standards, i.e., the concentration of contaminants like hydrocarbons and heavy metals is sufficiently low. Because such discharges are regulated, permits are required in most areas.

Offshore Discharges

Offshore discharges of treated solids, such as drill cuttings and produced solids, are permitted in some areas. Offshore discharges, however, are prohibited within three miles of shore in the United States, and the discharge of oil-based drilling mud wastes are prohibited in all United States waters. Where offshore discharges are prohibited, waste solids must be transported to shore for disposal (Arnhus and Slora, 1991). This is generally more expensive than offshore treating and discharge.

Onshore Discharges

Many solid wastes, particularly drill cuttings and produced solids, can be discharged by spreading them over the land surface. If the solids have been treated and are not contaminated with hydrocarbons, salt, or heavy metals, then obtaining permits for surface disposal may be relatively simple.

The suitability of a solid waste for surface discharge can be assessed through its electrical conductivity (EC), sodium adsorption ratio (SAR), the exchangeable sodium percentage (ESP), and the oil and grease (O&G) levels. Maximum · values generally recommended for these parameters are: EC < 4 mmhos/cm, SAR < 12, ESP < 15%, and O&G < 1% (Deuel, 1990). These parameters are discussed in more detail in Chapter 3. Another measure of the suitability of a solid waste for surface discharge is its heavy metal content. Maximum recommended accumulations of heavy metals in soil are presented in Table 7-1.

Treated waste solids can be used for road and site construction. Construction grade gravel and sand can be used as fill material on roads and drilling pads. Such use of treated solids minimizes the need for quarried gravel, which further lowers the environmental impact of drilling and production activities (Schumacher et al., 1990).

Land treatment can be used for the disposal of solids containing only hydrocarbons, particularly if the treatment is designed to degrade the hydrocarbons by biological processes (Bleckmann et al., 1989; Biederbeck, 1990). There are two major forms of land treatment in use: *landspreading* and *landfarming*. Landspreading is when wastes are spread over the surface of the ground and then tilled into the soil. After this initial tilling, no further action is usually taken. Landfarming is an enhanced version of landspreading in which additional processing of the soil is conducted after the initial tilling. In landfarming, the soil is commonly processed for several years after the initial application

Table 7-1
Maximum Recommended Heavy
Metal Concentration in Soil

Element	Soil Concentration (mg/kg)
Arsenic	300
Boron	3.*
Barium	**
Beryllium	50
Cadmium	3
Cobalt	200
Chromium	1,000
Copper	250
Mercury	10
Manganese	1,000
Molybdenum	5
Nickel	100
Lead	1,000
Selenium	5
Vanadium	500
Zinc	500

Concentration in soil-paste extract.
**Depending on site conditions, can be as high as 100,000 mg/kg.*

Source: Anderson et al., 1983.
Copyright Butterworth-Heinemann Publishers, 1983, with permission.

of the waste solids. This additional processing may include adding fertilizers and tilling repeatedly to increase oxygen uptake in the soil.

Most farmers do not object to landspreading because it provides some irrigation, helps condition the soil, stabilizes wind erosion, improves soil structure, and can improve crop yield (American Petroleum Institute, 1983; Deuel, 1990; Zimmerman and Robert, 1990).

There are two significant problems with land treatment that may limit future applications. First, land treatment provides little control over where mobile (leachable) fractions of the waste will go. Second, the spreading of oily wastes results in emissions of volatile organic compounds. These problems may result in a treatment project to be in violation of some applicable laws and regulations governing air pollution. This has led to land treatment being banned in some areas.

Road spreading is another disposal method for hydrocarbon-contaminated solids. The wastes are mixed with other construction materials and spread over gravel roads. The oil helps hold the road materials together, making such wastes an effective dust suppressant. Depending on the quality and type of solids, these wastes may also be used in the construction of new road beds. Road spreading is commonly used for the disposal of produced sand in Alberta and has been tested in California. Not all wastes are suitable for road spreading, however. The waste must not contain significant amounts of salt water, fracturing acids, other nonhydrocarbon contaminants, halogenated hydrocarbons, or manufactured oils. The hydrocarbons must be nonvolatile to minimize air pollution problems. Produced sand from heavy oil operations are well suited for road spreading because of their low content of aromatics and volatile hydrocarbons.

The environmental impact of road spreading is low for properly prepared wastes. The metals content of most oily wastes can be lower than that of asphalt, a common road paving material. Elevated levels of chloride, metals, or hydrocarbons have not been observed in ditch samples collected along roads used for the disposal of solid wastes (Kennedy et al., 1990; Cornwell, 1993). Because most of the wastes that are candidates for road spreading are high-volume, low-toxicity solids, disposal by road spreading reduces the volume of wastes that must be disposed of in overused landfills. Nevertheless, the lack of control over the spread of wastes is expected to limit and may even prohibit its future use.

7.2 SUBSURFACE DISPOSAL

Subsurface disposal is the most widely used method for the disposal of most petroleum industry wastes. Liquids are usually injected into deep subsurface formations through injection wells, and solids are usually buried in shallow pits at a drill site. If wastes are considered hazardous under applicable regulations, however, disposal at a licensed hazardous waste disposal site may be required.

7.2.1 Disposal of Liquids

The most common disposal method for waste liquids, such as produced water, is to inject them into a subsurface formation. Details on planning, installation, operation, and maintenance of disposal wells have been provided by the American Petroleum Institute (1978). The cost of drilling and completing a disposal well can be a significant expense in wastewater disposal.

Disposal wells must be completed in a formation that is permeable and porous, and has a low pressure and a large storage volume. The disposal formation must also be geologically isolated from any freshwater aquifers. To prevent the water from plugging the formation, the water must normally be treated to remove free and emulsified oil, suspended solids, and some dissolved solids, such as iron and scale, prior to disposal.

One disposal method that is growing in popularity is annular injection in existing wells. In this process, the wastewater is injected down the annulus of an existing injection or production well and into a formation above the existing completion. A packer is used to isolate the disposal zone from the existing injection or production zone. This disposal method can eliminate the cost of drilling a separate disposal well. The disposal zone must still meet all requirements for protecting freshwater aquifers.

A major concern with underground disposal of water is the potential for the well to provide a vertical communication path from the disposal formation to any overlying freshwater aquifers. Possible communication paths include flow up the inside of the casing through leaks in the casing and flow up the outside of the casing through a bad cement bond. The presence of leaks in the casing can be detected through mechanical integrity tests. Unfortunately, there are no reliable ways to detect the flow of water up the outside of the casing.

Mechanical integrity tests are required in the United States to determine whether leaks are present in casing. These tests are conducted with tubing set in casing. Two types of tests are commonly used (Kamath, 1989). In one type of test, the level of any liquid in the annulus above the packer is monitored for changes. In most cases, the fluid level will rise as fluid leaks from the higher-pressured disposal zone to the lower-pressured zones overlying it. In the second type of test, the annulus is pressurized and its pressure is then monitored. If there is a leak, the pressure in the annulus will decline. The annulus pressure method, however, requires that the well be isothermal and that there are no interactions between the liquid and gas in the annulus. Because these requirements are rarely present, the fluid level method is normally considered more reliable than the annulus pressure method for detecting leaks in the casing.

Although there are no reliable methods to detect whether or not water is flowing up the outside of casing, a number of methods are available that can detect flow in some cases. These methods are generally limited to large leaks or high fluid flow rates. Noise logs can detect high-volume fluid movement behind casing, but are sensitive to extraneous sources of sound. Neutron activation logs can detect water movement in some cases (Arnold and Paap, 1979; McKeon et al., 1991; Uswak and Howes, 1992). Injecting a boron solution into the well and logging with a pulsed neutron log to monitor boron migration has been proposed (Bount et al., 1991). Radioactive tracers can be injected into the well and the well logged with a gamma-ray detector to test for tracer movement. Acoustic cement bond logs can provide evidence of a bad cement bond in some cases.

The possibility of pressure communication between the disposal zone and overlying formations can be tested by separately completing an intermediate zone above the disposal zone and below the overlying freshwater aquifer. If the pressure in the intermediate zone responds to the injection pressure into the disposal zone, then a leak behind the casing is indicated (Poimboeuf, 1990).

If a well fails a mechanical integrity test, the well normally must be repaired before it can be used as an injection well. One method of repairing a leak in casing is to install a concentric packer to isolate the leak and allow fluid flow past the bad section of casing (Wilson, 1990). Other methods to repair wells that fail a mechanical integrity test include squeeze cementing, running a liner, or plugging and abandoning the well.

Failure of a mechanical integrity test does not necessarily mean that freshwater aquifers will be contaminated; it only indicates the possibility of water flow up the annulus. No upflow will occur if the disposal formation is underpressured, e.g., if its flow potential (absolute pressure minus hydrostatic gradient) is lower than that of the overlying zones. It has been suggested that wells that fail a mechanical integrity test and are underpressured could still be safely used without repair if the fluid level in the annulus is continuously monitored to ensure that no vertical flow of fluids occurrs (Janson and Wilson, 1990). The disposal of water by gravity feed, by which there are no pumps to pressurize the disposal zone, has also been proposed for such wells (Meyer, 1990). Regulatory approval to use a well that fails a mechanical integrity test may not be possible in some areas.

7.2.2   Disposal of Solids

Subsurface burial is a common method for disposal of solid wastes. Drill cuttings and used mud are typically left in reserves pit after a well is drilled. After the free liquids are removed, the remaining materials are covered by soil and the site is revegetated. Such on-site disposal is allowed in most areas, provided there are no hazardous materials mixed with the waste.

One major concern with the burial of solids is the potential for heavy metals, hydrocarbons, and salts to migrate away from the site. Salt buried in reserves pits can migrate both downward and upward (McFarland et al., 1990). The two metals most commonly found in drilling muds at concentrations above those found in most soils are barium and chromium. These metals, along with mercury, are in a nonsoluble form and have a very limited potential for migration or plant uptake (American Petroleum Institute, 1983). For pits containing high salt or hydrocarbon levels, regulations may require the use of an impermeable pit liner to prevent leaching. The leaching rate for unlined pits could also be reduced by covering the buried waste with an impermeable cap to prevent stormwater infiltration (Roberts and Johnson, 1990).

In arctic regions, the disposal of drilling wastes in pits using below-grade freezeback has been proposed. In this process, the drilling wastes are buried in a deep pit dug into the permafrost. After closure, the materials will freeze, minimizing any migration of soluble components from the site. Only water-based muds, cuttings, and excess cement can be successfully disposed of by this method; any freeze-depressing materials like brines, glycols, or alcohols may be prohibited (Maunder et al., 1990). The long-term stability of these pits in the event of climate changes, however, is not known (Fristoe, 1990).

A developing new technology for the disposal of drill cuttings is to grind them into small particles and inject them into a well as a slurry (Malachosky et al., 1991; Smith, 1991; Minton and Secoy, 1993). In most cases, annular injection is used for the slurry. If fracturing is required for the slurry to be accepted by the formation, it will be necessary to ensure that the disposal zone and any hydraulic fractures remain isolated from overlying freshwater aquifers (Andersen et al., 1993). The design of slurry injection projects can be difficult, however, because reliable data on the rheology and fracturing properties of the slurry are limited (Crawford and Lescarboura, 1993).

If hazardous materials are present, regulations may require that the wastes be shipped to a commercial off-site disposal facility. Materials that normally cannot be disposed of by on-site burial include pipe dope cans, waste lubricating oils, mud sacks, solvents, or excess treatment chemicals. In most cases, commercial off-site disposal facilities consist of an engineered landfill. If the landfill is permitted to accept hazardous wastes, it must have a synthetic liner with a leachate monitoring and collection system. Other types of landfills, e.g., those with clay liners and that have less stringent monitoring requirements, can accept nonhazardous wastes. Waste disposal at commercial facilities should be used with caution, however, because hazardous waste regulations in the United States can impose liability on all companies for any wastes at the facility, regardless of who actually sent any particular waste.

Naturally occurring radioactive materials (NORM) generated at production sites must also be safely disposed of in ways to prevent unnecessary human exposure to nuclear radiation. Several studies have concluded that many disposal methods are available that are effective in keeping human exposure to nuclear radiation from NORM well below 100 mREM/year (American Petroleum Institute, 1990; Miller and Bruce, 1990). These disposal methods included landspreading, landspreading with dilution, shallow burial, disposal in plugged and abandoned wells, and subsurface injection (with or without hydraulic fracturing). Regulations governing the disposal of NORM are currently being formulated. Until approved disposal options become available, NORM contaminated equipment and soil should be stored on-site.

Abandoned offshore platforms must also be disposed of. The platform must be removed to eliminate any navigational hazards it poses. In most instances, explosives are used for cutting the legs to free the platform from the sea floor. Such explosive cutting has been identified as a possible cause of deaths of endangered sea turtles and marine animals (Arscott, 1989). Other methods of cutting platform legs have been considered, including sawing with diamond wires, flame cutting with acetylene and oxygen, arc cutting with steel electrodes, plasma arc cutting with argon, cryogenic fragmentation, and high-pressure water jet cutting . These methods, however, are significantly more hazardous to the work crews involved (McNally, R., 1987; Al-Hassani, 1988; Pittman et al., 1961; Murrell and Faul, 1989).

Once the platform has been removed, it can be disposed of by transporting it to land, cutting it into pieces, and burying it. This process, however, is expensive. In some areas, abandoned platforms can be sunk to the sea floor and used as artificial reefs to enhance offshore fisheries. The platform provides a solid substrate for aquatic plants to grow, which then attract fish. The "Rigs to Reefs" program may be particularly attractive in offshore areas having few natural reefs, such as the U.S. Gulf of Mexico. Such programs are currently being developed in a number of states.

REFERENCES

Al-Hassani, S. T., S., "Platform Removal Demands complex Explosive Designs," *Oil and Gas J.,* May 16, 1988.

American Petroleum Institute, *Subsurface Salt Water Injection and Disposal,* Book 3 of the API Vocational Training Series, Washington, D.C., 1978.

American Petroleum Institute, "Summary and Analysis of API Onshore Drilling Mud and Produced Water Environmental Studies," API Bulletin D19, Washington, D.C., Nov. 1983.

American Petroleum Institute, "Management and Disposal Alternatives for NORM Wastes in Oil Production and Gas Plant Equipment," Washington, D.C., May 1990.

Anderson, D.C., Smith, C., Jones, S. G., and Browh, K. W., "Fate of Constituents in the Soil Environment, in Hazardous Waste Land Treatment," K. W. Brown, G. B. Evans, Jr., and B. D. Frentrup (editors), Wobern, MA: Butterworth Publishers, 1983.

Andersen, E. E., Louviere, R. J., and Witt, D. E., "Guidelines for Designing Safe, Environmentally Acceptable Downhole Injection Operations," paper 25964 presented at the Society of Petroleum Engineers/Environmental Protection Agency's Exploration and Production Environmental Conference, San Antonio, TX, March 7–10, 1993.

Arnhus, K. M. and Slora, G., "Cuttings and Waste Mud Disposal," paper SPE/IADC 21949 presented at the Society of Petroleum Engineers SPE/IADC Drilling Conference, Amsterdam, The Netherlands, March 11–14, 1991.

Arnold, D. M. and Paap, H. J., "Quantitative Monitoring of Water Flow Behind and in Wellbore Casing," *J. Pet. Tech.,* Jan. 1979, pp. 121–130.

Arscott, R. L., "New Directions in Environmental Protection in Oil and Gas Operations," *J. Pet. Tech.,* April 1989, pp. 336–342.

Biederbeck, V. O., "Using Oily Waste Sludge Disposal to Conserve and Improve Sandy Cultivated Soils," Proceedings of the U.S. Environmental Protection Agency's First International Symposium on Oil and Gas Exploration and Production Waste Management Practices, New Orleans, LA, Sept. 10–13, 1990, pp. 1025–1038.

Bleckmann, C. A., Gawel, L. J., Whitfill, D. L., and Swindoll, C. M., "Land Treatment of Oil-Based Drill Cuttings," paper SPE 18685 presented at the 1989 Society of Petroleum Engineers SPE/IADC Drilling Conference, New Orleans, LA, Feb. 28–March 3, 1989.

Bount, C. G., Copoulos, A. E., Myers, G. D., "A Cement Channel-Detection Technique Using the Pulsed-Neutron Log," *SPE Formation Evaluation*, Dec. 1991, pp. 485–492.

Cornwell, J. R., "Road Mixing Sand Produced from SteamDrive Operations," paper SPE 25930 presented at the Society of Petroleum Engineers/Environmental Protection Agency's Exploration and Production Environmental Conference, San Antonio, TX, March 7–10, 1993.

Crawford, H. R. and Lescarboura, J. A., "Drill Cuttings Reinjection for Heidrun: A Study," paper SPE 26382 presented at the Society of Petroleum Engineers 68th Annual Technical Conference and Exhibition, Houston, TX, Oct. 3–6, 1993.

Deuel, L. E., "Evaluation of Limiting Constituents Suggested for Land Disposal of Exploration and Production Wastes," Proceedings of the U.S. Environmental Protection Agency's First International Symposium on Oil and Gas Exploration and Production Waste Management Practices, New Orleans, LA, Sept. 10–13, 1990, pp. 411–430.

Fristoe, B., "Drilling Wastes Management for Alaska's North Slope," Proceedings of the U.S. Environmental Protection Agency's First International Symposium on Oil and Gas Exploration and Production Waste Management Practices, New Orleans, LA, Sept. 10–13, 1990, pp. 281–292.

Janson, L. G., Jr. and Wilson, E. M., "Application of the Continuous Annular Monitoring Concept to Prevent Groundwater Contamination by Class II Injection Wells," Proceedings of the U.S. Environmental Protection Agency's First International Symposium on Oil and Gas Exploration and Production Waste Management Practices, New Orleans, LA, Sept. 10–13, 1990, pp. 73–92.

Kamath, K. I., "Regulatory Control of Groundwater Contamination by Hazardous Waste Disposal Wells: An Engineering Perspective," paper SPE 19744 presented at the Society of Petroleum Engineers 64th Annual Technical Conference and Exhibition, San Antonio, TX, Oct. 8–11, 1989.

Kennedy, A. J., Holland, L. L., and Price, D. H., "Oil Waste Road Application Practices at the Esso Resources Canada Ltd., Cold Lake Production Project," Proceedings of the U.S. Environmental Protection Agency's First

International Symposium on Oil and Gas Exploration and Production Waste Management Practices, New Orleans, LA, Sept. 10–13, 1990, pp. 689–702.

Malachosky, E., Shannon. B. E., and Jackson, J. E., "Offshore Disposal of Oil-Based Drilling Fluid Waste: An Environmentally Acceptable Solution," paper SPE 23373 presented at the Society of Petroleum Engineers First International Conference on Health, Safety, and Environment, The Hague, Netherlands, Nov. 10–14, 1991.

Maunder, T. E., Le, K. M., and Miller, D. L., "Drilling Waste Disposal in the Arctic Using Below-Grade Freezeback," paper SPE 20429 presented at the Society of Petroleum Engineers 65th Annual Technical Conference and Exhibition, New Orleans, LA, Sept. 23–26, 1990.

McFarland, M., Ueckert, D. N., and Hartmann, S., "Evaluation of Selective-Placement Burial for Disposal of Drilling Fluids in West Texas," Proceedings of the U.S. Environmental Protection Agency's First International Symposium on Oil and Gas Exploration and Production Waste Management Practices, New Orleans, LA, Sept. 10–13, 1990, pp. 455–466.

McKeon, D.C., Scott, H. D., Olesen, J. R., Patton, G. L., and Mitchell, R. J., "Improved Oxygen-Activation Method for Determining Water Flow Behind Casing," *SPE Formation Evaluation,* Sept. 1991, pp. 334–342.

McNally, R., "Variety of Factors Impact Platform Removal," *Petroleum Engineer International,* April 1987.

Meyer, L., "Simple Injectivity Test and Monitoring Plan for Brine Disposal Wells Operating by Gravity Flow," Proceedings of the U.S. Environmental Protection Agency's First International Symposium on Oil and Gas Exploration and Production Waste Management Practices, New Orleans, LA, Sept. 10–13, 1990, pp. 865–872.

Miller, H. T. and Bruce, E. D., "Pathway Exposure Analysis and the Identification of Waste Disposal Options for Petroleum Production Wastes Containing Naturally Occurring Radioactive Materials," Proceedings of the U.S. Environmental Protection Agency's First International Symposium on Oil and Gas Exploration and Production Waste Management Practices, New Orleans, LA, Sept. 10–13, 1990, pp. 731–744.

Minton, R. C. and Secoy, B., "Annular Reinjection of Drilling Wastes," *J. Pet. Tech.,* Nov. 1993, pp. 1081–1085.

Murrell, D. and Faul, R., "Platform Removal with High-Pressure Fluids Environmentally Sound and Efficient," Proceedings of the Petro-Safe '89 Conference, Houston, TX, Oct. 3–5, 1989.

Pittman, F. C., Harriman, D. W., and St. John, J. C., "Investigation of Abrasive-Laden-Fluid Method for Perforation and Fracture Initiation," *J. Pet. Tech.,* May 1961, pp. 489–495.

Poimboeuf, W. W., "An Early Warning System to Prevent USDW Contamination. Environmental Underground Injection Equipment for Hazardous and Non-Hazardous Liquid Waste Disposal. Injection Well and Monitoring Well in the Same Borehole," Proceedings of the U.S. Environmental Protection Agency's First International Symposium on Oil and Gas Exploration and Production Waste Management Practices, New Orleans, LA, Sept. 10–13, 1990, pp. 43–72.

Roberts, L. and Johnson, G., "A Study of the Leachate Characteristics of Salt Contaminated Drilling Wastes Treated with a Chemical Fixation/ Solidification Process," Proceedings of the U.S. Environmental Protection Agency's the First International Symposium on Oil and Gas Exploration and Production Waste Management Practices, New Orleans, LA, Sept. 10–13, 1990, pp. 933–944.

Schumacher, J. P., Malachosky, E., Lantero, D. M., and Hampton, P. D., Minimization and Recycling of Drill Cuttings for the Alaskan North Slope," paper SPE 20428 presented at the Society of Petroleum Engineers 65th Annual Technical Conference and Exhibition, New Orleans, LA, Sept. 23–25, 1990.

Smith, R. I., "The Cuttings Grinder," paper SPE 22092 presented at the Society of Petroleum Engineers International Arctic Technical Symposium, Anchorage, AK, May 29–31, 1991.

Uswak, G. and Howes, E., "Direct Detection of Water Flow Behind Pipe Using a Transient Oxygen Activation Technique," *Journal of Canadian Petroleum Technology,* Vol. 31, No. 4, April 1992, pp. 38–45.

Wilson, E. M., "'The Application of Concentric Packers to Achieve Mechanical Integrity for Class II Wells in Osage County, Oklahoma," Proceedings of the U.S. Environmental Protection Agency's First International Symposium on Oil and Gas Exploration and Production Waste Management Practices, New Orleans, LA, Sept. 10–13, 1990, pp. 967–976.

Zimmerman, P. K., and Robert, J. D., "Landfarming Oil Based Drill Cuttings," Proceedings of the U.S. Environmental Protection Agency's First International Symposium on Oil and Gas Exploration and Production Waste Management Practices, New Orleans, LA, Sept. 10–13, 1990, pp. 565–576.

CHAPTER 8

Remediation of Contaminated Sites

Many petroleum industry sites have been contaminated from previous activities or can become contaminated through accidental releases of various materials. In many cases, remediation will be required to restore the impacted area. Sites that may require remediation include old reserves pits, onshore release sites of hydrocarbons or contaminated water, and places where oil slicks from offshore releases are blown onshore.

Unfortunately, the complex pore structure and fluid transport pathways of soil can make remediation difficult. Hydrocarbons can be found in various places in soil. Most are trapped by capillary pressure as a discrete liquid phase within the pores of the soil. If a sufficient volume of hydrocarbons has been released, it can exist in a separate, mobile phase that floats on top of the groundwater. Hydrocarbons can also be dissolved into the groundwater and be transported away from the release site. Volatile hydrocarbons can be found as a vapor in air-saturated pores. Dissolved solids can also be found in various places in soil. They can remain dissolved and migrate with groundwater or they can be absorbed onto the soil solids.

Because of the complex distribution of contaminants in soil, a comprehensive site evaluation may be required before the optimum remediation process can be selected and properly implemented. A number of site remediation methods are available and are reviewed below.

8.1 SITE ASSESSMENT

An important process in the cleanup of contaminated sites is to assess their potential to impact human health and the environment before remediation begins. A site assessment is the first step in determining what remediation method is to be used, if any.

Site assessments are normally conducted in stages (American Petroleum Institute, 1989). The first stage is to gather all relevant background information about the site. This includes reviewing available records and reports and may include interviewing site personnel. From this information, the magnitude and composition of the release is estimated. The next stage is to characterize the site. The purpose of site characterization is to determine the exact locations, contaminant concentrations, and extent of the contaminated zone and to evaluate the potential for the contaminants to migrate from the site. Depending on the magnitude of the release and its potential for adverse environmental impact, a risk assessment study may be needed. A risk assessment study would quantify the potential environmental impact of the various remediation processes being considered, and the results could be used when negotiating the specific details of a site remediation project with applicable regulatory agencies.

An important part of site assessment is to develop a detailed sampling and sample analysis plan that clearly identifies the objectives of the analysis and how those objectives are to be met. This plan must also address quality assurance and control to ensure that the data obtained accurately reflect the actual concentrations being measured (Keith et al., 1983). This plan determines the number of samples to be obtained and their locations. Normally, a network of sampling points is selected around the site. Geological and hydrological factors must be considered in selecting each point, including any local groundwater flow, the hydraulic conductivity (permeability) of the soil, geological heterogeneities that can affect fluid flow, and geochemical processes, such as ion exchange, that can cause contamination to migrate at a rate different from the physical flow of groundwater. Additional sampling sites may be added if a statistical analysis of the initial samples indicates that the confidence limits are unacceptably low (Wojtanowicz et al., 1989).

Two types of samples are generally obtained: core samples and fluid samples. Core samples provide information about liquids trapped with the solids by capillary pressure, while fluid samples provide information about mobile liquids. The type of samples required depends on the type of contaminant and may be specified by applicable regulatory agencies. Other types of measurements, such as electromagnetic surveys to characterize the extent of brine plumes, can also be used (Young, 1991; Dalton, 1993). A detailed discussion of sampling and analysis is given by Johnson and James (1989).

Sampling procedures must be designed to provide unbiased data throughout the cleanup process (American Petroleum Institute, 1983 and 1985a). Procedures that can contaminate samples include how the sampling device is emplaced at the site. Drilling can alter the in situ geochemical environment by flushing with drilling fluid or by allowing the commingling of fluids in different zones. The presence of foreign materials like grout or bentonite around the wellbore can contaminate water flowing into the well. To flush such contaminants from the well, typically eight to ten well volumes of water are pumped from the well before samples are taken. Samples can also become contaminated from exposure to atmospheric pollutants, particularly oxygen. The composition of obtained samples can be altered by degassing or by sitting stagnant for a few weeks before testing, allowing bacteria to grow. Finally, human error in any of the stages of obtaining the sample can result in sample contamination.

For groundwater remediation projects that are expected to take a number of years to complete, the timing of sampling during remediation should be systematic, not random. Systematic sampling is easier to schedule and administer and will allow seasonal variations to be identified and accounted for. For many cleanup projects, sampling four to six times per year may be adequate to ensure that the cleanup dynamics are observed and to minimize expensive redundancy (Johnson and Jennings, 1990).

Once obtained, the samples must then be accurately analyzed. A variety of analytical procedures are available for different contaminants (National Research Council, 1985; American Petroleum Institute, 1985g and 1987a). For compounds with concentrations in the parts-per-million or higher range, the accuracies of most analytical procedures are generally good. For trace contaminants, however, consistent results may be difficult to obtain. Different measurement methods can also yield different results, particularly if improper extraction methods are used (Holliday and Deuel, 1993). Sophisticated analytical techniques like gas chromatography may be required for accurate results. Regulatory agencies may specify the types of analyses that must be conducted.

After the data have been obtained, a statistical analysis of the data must be conducted. The type of statistical analysis conducted will depend on the needs of the study and how the resulting conclusions will be used. A variety of statistical tools are available through time series and trend analysis. When analyzing data, however, it is important to include all of the data in the analysis. Using the lowest measured concentrations and discarding the highest may be illegal.

The difficulty in obtaining reliable data for the composition of contaminated sites was dramatically demonstrated through independent studies conducted by the U.S. Environmental Protection Agency and the American Petroleum Institute (Holliday and Deuel, 1990). In these studies, the same reserves pits, produced water pits, and production facilities were independently sampled and analyzed by the two agencies. As seen in Table 8-1, no correlation was found between the

Table 8-1
Correlation of Independent Measurements of Waste Composition

Constituent	Pit Liquids	Pit Solids	Produced Water
Sodium	Good	Poor	Good
Potassium	Good	None	Good
Calcium	Good	Good	Good
Magnesium	Good	None	Good
Sum of cations	Good	—	Good
Chlorides	Poor	None	Good
Electrical conductivity	—	—	None
Sodium absorption ratio	—	—	Good
Total suspended solids	None	—	None
Arsenic	None	None	None
Boron	Poor	None	Good
Barium	None	Good	Good
Chromium	—	Poor	—
Copper	—	Good	—
Lead	—	None	—
Nickel	—	None	—
Zinc	—	Poor	—
Oil and grease	Good	None	Poor
Total organic carbon	None	Poor	Good
Napthalene	None	None	None
Toluene	None	None	Good
Ethylbenzene	—	None	None
pH	Good	None	None
Moisture content	—	None	—

Source: after Holliday and Deuel, 1990.
Copyright SPE, with permission.

concentrations obtained by the two agencies for many important waste constituents. Further, the correlation was not improved by sending samples to different laboratories. Unfortunately, insufficient data were obtained to determine the exact cause of the poor correlation.

An important step in any cleanup program is to determine when cleanup is complete. This includes determining the acceptable level of residual contaminants. Acceptable levels are typically determined by comparing the contaminant levels to the standards for "clean" water or to levels that existed prior to the release. These levels are normally site-specific and are determined by negotiation with the appropriate regulatory agencies (Cooper and Hanson, 1990). Determining cleanup levels may require risk assessment studies to be conducted.

Because of the statistical variability in any data and possible problems with sampling and analysis, identifying when a particular standard has been achieved can be difficult. Remediation to where the applicable standard is met with a 90% confidence limit is often used, although the actual level required is normally determined through negotiation with applicable regulatory agencies on a case-by-case basis (Hoffman, 1993).

In determining whether further remediation is required, political and institutional pressures that have no relevance to protecting human health or the environment often exist. Too often these pressures are not based on accurate scientific information and can result in additional remediation costs with little benefit to the environment. This problem is further compounded by the disagreement on what scientific standards should be used. Even the relatively simple question of whether cleanup should be based on the level of total petroleum hydrocarbons or the levels of individual hydrocarbon compounds has not been resolved. In some cases, the actual risk to the environment often depends more on the composition of a contaminant than on its total amount, particularly when only trace quantities are present. The importance of accurate scientific information is evidenced in the conflicting stories about the environmental impact related to the Exxon Valdez spill (Maki et al., 1993).

8.2 REMEDIATION PROCESSES

A number of remediation processes are available to clean up contaminated sites (Sims, 1990). Some are suitable for cleaning up contaminated soil or subsurface groundwater, while others are suitable for cleaning up contaminated shorelines. These processes vary significantly in how completely they remove the contaminants, in the time they require, and in their associated costs. Successful cleanup may require a combination of remediation processes (Peters and Wentz, 1991). Cleanup of offshore oil releases is discussed in Appendix D.

Because most remediation processes entail their own environmental hazards, care must be taken in selecting which method to use. The potential impact will depend on the remediation process used and the type of habitat being cleaned. Shoreline habitats are particularly sensitive to remediation processes (American Petroleum Institute, 1985d and 1985e).

One of the concerns that must be addressed when designing remediation projects is to prevent any further spread of the contaminant plume. Containment of the plume is of particular concern if remediation is expected to take a number of years, as it may for contaminated groundwater. Containment can be accomplished by placing physical or hydraulic barriers around it. Physical barriers consist of an impermeable material that is emplaced around the contaminant plume to prevent its migration. Grout can be injected into the soil, where it solidifies to form an impermeable barrier. Sheet piling made of steel plates can be driven into the ground around the contaminant. Trenches can be dug and backfilled with an impermeable medium to form a slurry wall. Hydraulic barriers consist of a set of wells around the site from which fluid is withdrawn at a rate at least equal to the groundwater flow rate. The withdrawal point becomes a low point for the hydraulic pressure, inducing all groundwater in the immediate area, including the contaminant plume, to flow to the wells instead of away from the site.

Natural Processes

For some contaminated sites, the best remediation process may be to do nothing and let natural processes degrade and remove any contaminants. This option may be particularly suitable for oil spills in sensitive shoreline habitats, where implementing a remediation process may cause more damage than leaving the spilled oil (Kiesling et al., 1988). Natural processes that remove hydrocarbons include evaporation, photo-oxidation, and bacterial action, coupled with dispersion from wind and wave action. Natural processes can be enhanced by manually removing free oil with absorbants.

Pump and Treat

Traditionally, the most common way to remediate hydrocarbon-contaminated groundwater is to pump the groundwater from on-site wells or trenches and then treat the water. Pumping creates a fluid cone or water depression around the well that establishes a hydraulic gradient to drive fluid to the well faster than the normal groundwater flow rate. Because of this, pump and treat normally produces large volumes of water. The produced water must be treated to remove the hydrocarbons before being discharged. Treated water can then be reinjected for disposal or to help drive remaining contaminants to the pumping wells. Treatment methods for hydrocarbon-contaminated water are discussed in Chapter 6.

One proposed method to minimize the volume of water produced is to install pumps with oil-wet elements that only allow entry of hydrocarbons. These pumps, however, do not establish the water depression around the wellbore, limiting the flow rate of hydrocarbons to the well. The use of horizontal wells to improve hydrocarbon capture has also been proposed (Karisson, 1993).

Because of capillary trapping of hydrocarbons in the pore spaces, pump and treat will not completely recover all of the hydrocarbons at a spill site. These trapped hydrocarbons may be water soluble and can dissolve into the groundwater as it flows past. Thus, trapped hydrocarbons can provide a source of groundwater contamination long after all free hydrocarbons have been removed. Because of this, additional remediation may be required.

Soil Flushing

One way to speed the removal of hydrocarbons from soil is to flush water through the spill site. If additives are used with the water, many of the trapped hydrocarbons can also be removed. The environmental impacts of any additives used, however, must be low.

Surfactants and other chemicals can be added to the water to enhance the removal of trapped hydrocarbons in soil. These materials lower the capillary pressure between the hydrocarbons and groundwater, allowing the hydrocarbons to migrate through the pore structure to the wells, where they can be removed. Chemicals can also increase the solubility of some hydrocarbon compounds in water. This increased solubility can accelerate the removal of trapped hydrocarbons by continued flushing. Laboratory studies of gasoline recovery using surfactants have been successful (American Petroleum Institute, 1985f and 1986g).

Flushing can also enhance remediation of hydrocarbon-contaminated shorelines. For example, small barriers and channels can be constructed to enhance natural flushing in tidal flats or river deltas to remove oil from stagnant areas. Oil on rocky beaches can be hosed off by jets of water or steam. High-pressure or high-temperature jets can be more effective in removing the hydrocarbons, but can result in more damage to the habitat.

Vaporization

Volatile hydrocarbons can be removed from soil by vaporization. Natural vaporization can be enhanced by tilling the soil. For hydrocarbons located deeper than normal tilling depths, vaporization can be enhanced by injecting air or by pulling a vacuum on the soil. Air injection has proven effective in removing gasoline in both laboratory and field studies (American Petroleum Institute, 1985b and 1986b). This process lowers the partial pressure of the hydrocarbon in the vapor phase in the soil, inducing further vaporization. Air injection is generally more effective at hydrocarbon removal, although vacuum extraction requires less air to be handled.

An emerging variation on volatilization is to heat the soil. Because the vapor pressure of volatile hydrocarbons increases almost exponentially with temperature, volatilization can be significantly enhanced through heating. Injecting steam has proven to be effective in vaporization of volatile hydrocarbons (Hunt et al., 1988; Udell and Stewart, 1990). Heating through radio frequency or electrical currents has also been proposed.

Volatilization may not be a good remediation process if the hydrocarbon contaminant contains nonvolatile components. Once the volatile components have been removed, the remaining components will be heavier, more viscous, and less likely to be recovered by any subsequent processes. However, because the most toxic hydrocarbon components tend to be the most volatile, any remaining hydrocarbons in the soil would tend to have a relatively low toxicity. Because volatilization releases materials that may contribute to air pollution, permits may be required.

Bioremediation

Biological degradation can be used at some sites to remove hydrocarbons in soil. Fertile soil naturally contains up to one million hydrocarbon-degrading bacteria per gram of dry soil (Testa and Winegardner, 1991). By adding nutrients and ensuring the availability of oxygen, in situ bioremediation can effectively degrade many hydrocarbon contaminants. This process can take several months to several years to complete, however, and is difficult to control (U.S. Environmental Protection Agency, 1990; American Petroleum Institute, 1986a and 1986h).

One factor controlling the effectiveness of in situ bioremediation is the soil structure. Sandy soils with a high permeability allow higher levels of biological activity than do soils containing significant quantities of silt or clay. The more permeable soils permit a more rapid transport of air (oxygen), water, and nutrients to the sites of biological activity.

The availability of oxygen is another factor controlling the effectiveness of in situ bioremediation. To help ensure an adequate supply of oxygen, air is commonly injected into the formation in a process called air sparging. The injection of hydrogen peroxide has also been suggested as a means of increasing the oxygen levels, but its effectiveness has not been established (American Petroleum Institute, 1985c and 1986c). Hydrogen peroxide, however, is toxic and its use may not be permitted.

In situ bioremediation may not be allowed in some areas. Some regulatory agencies prohibit the injection of chemicals into groundwater, preventing the addition of nutrients needed for bacterial growth.

Excavation

For small sites, all of the contaminated soil can be excavated. The excavated soil can then be treated by one of the methods discussed in Chapter 6. The primary benefit of excavation is the insurance that all of the contaminant has been removed, which lowers the potential for any future liability costs. The primary disadvantage of excavation, however, is its high cost. Excavation and disposal at an off-site hazardous waste disposal facility may be the only economically viable cleanup option for sites contaminated with heavy metals like chromium (Campbell and Akers, 1990).

Brine Contamination

The most common remediation process for brine-contaminated soil is to flush the soil with fresh water to leach the salt away. In many areas, natural processes like flooding and rainwater provide sufficient fresh water to remove the salt. If natural processes are inadequate, remediation can be enhanced through irrigation.

Because soil is a strong cation exchange medium, particularly when it has a high clay content, remediation by leaching can be a very slow process. Leaching can be enhanced through the addition of a cationic solution of calcium, such as gypsum (calcium sulfate) or calcium nitrate. Calcium, which has a lower impact on plant growth than sodium, replaces sodium in the exchange sites in the clays, allowing sodium to be leached away more rapidly. Field tests have shown that applying a calcium solution has been successful in revegetating some brine-contaminated soils. If the soil already contains a high concentration of sulfate ions, as in the case of a reserves pit site with barium sulfate (barite), the solubility of gypsum can be lowered, rendering it less effective for remediation (Hartmann et al., 1990; American Petroleum Institute, 1983).

For soils marginally contaminated with brine, one simple remediation process is to increase the soil's fertility. An increase in fertility may allow plants to be grown in the contaminated soil. Native grass mulch or aged manure can be disked into the top foot of soil, followed by leaching with water. This will provide additional fertilizer, as well as opening the pore structure for improved water and air transport. These sites can also be revegetated with salt-tolerant plants (Ueckert et al., 1990).

Sulfur Contamination

At some production sites, sulfur has been removed from sour natural gas and then stored at the site. At many sites, molten sulfur has been deposited on the soil to create a base pad upon which additional sulfur has been piled.

The first step in cleaning these sites generally consists of excavating the pad. The excavated sulfur is then broken into pieces to remove rocks, logs, and other such materials. Separation from the remaining soil can be achieved by either melting the sulfur or through a froth flotation process. In froth flotation, pieces of soil and sulfur are agitated in water to break them up. The mixture is then aerated, and the lighter sulfur particles attach to the air bubbles and float to the top where they are separated from the solids (Adamache, 1990).

The contaminated soil around the pad can be neutralized by adding calcium carbonate (limestone) to the soil. The soil structure, organic carbon content, and nutrient levels may also need to be restored. Reclamation of these sites may take five to seven years (Leggett and England, 1990).

REFERENCES

Adamache, I., "Contaminated Sulphur Recovery by Froth Flotation," Proceedings of the First International Symposium on Oil and Gas Exploration and Production Waste Management Practices, New Orleans, LA, Sept. 10–13, 1990, pp. 185–198.

American Petroleum Institute, "Groundwater Monitoring and Sample Bias," API Publication 4367, June 1983.

American Petroleum Institute, "Field Evaluation of Well Flushing Procedures," API Publication 4405, June 1985a.

American Petroleum Institute, "Subsurface Venting of Hydrocarbon Vapors from an Underground Aquifer," API Publication 4410, Sept. 1985b.

American Petroleum Institute, "Feasibility Studies on the Use of Hydrogen Peroxide to Enhance Microbial Degradation of Gasoline," API Publication 4389, May 1985c.

American Petroleum Institute, "Oil Spill Response: Options for Minimizing Adverse Ecological Impacts," API Publication 4398, Aug. 1985d.

American Petroleum Institute, "Oil Spill Cleanup: Options for Minimizing Adverse Ecological Impacts," API Publication 4435, Dec. 1985e.

American Petroleum Institute, "Test Results of Surfactant Enhanced Gasoline Recovery in a Large-Scale Model Aquifer," API Publication 4390, April 1985f.

American Petroleum Institute, "Detection of Hydrocarbons in Groundwater by Analysis of Shallow Soil Gas/Vapor," API Publication 4394, 1985g.

American Petroleum Institute, "Field Application of Subsurface Biodegradation of Gasoline in a Sand Formation," API Publication 4430, 1986a.

American Petroleum Institute, "Forced Venting to Remove Gasoline Vapor from a Large-Scale Model Aquifer," API Publication 4431, 1986b.

American Petroleum Institute, "Enhancing the Microbial Degradation of Underground Gasoline by Increasing Available Oxygen," API Publication 4428, 1986c.

American Petroleum Institute, "Underground Movement of Gasoline on Groundwater and Enhanced Recovery by Surfactants," API Publication 4317, 1986g.

American Petroleum Institute, "Beneficial Stimulation of Bacterial Activity in Ground Waters Containing Petroleum Products," API Publication 4427, 1986h.

American Petroleum Institute, "Manual of Sampling and Analytical Methods for Petroleum Hydrocarbons in Groundwater and Soil," API Publication 4449, 1987a.

American Petroleum Institute, "A Guide to the Assessment and Remediation of Underground Petroleum Releases," API Publication 1628, August 1989.

Campbell, R. E. and Akers, R. T., "Characterization and Cleanup of Chromium-Contaminated Soil for Compliance with CERCLA at the Naval Petroleum Reserve No. 1 (Elk Hills): A Case Study," paper SPE 20714 presented at the 65th Annual Technical Conference and Exhibition, New Orleans, LA, Sept. 23–26, 1990.

Cooper, D. E. and Hanson, J. B., "Establishing Site Specific Cleanup Standards at Superfund Sites," proceedings of How Clean is Clean? Cleanup Criteria for Contaminated Soil and Groundwater, Air and Waste Management Association, Boston, MA, Nov. 7–9, 1990.

Dalton, M. S., "The Use of Geophysical Surveys for Conductive Groundwater Plume Assessment During Remediation," paper SPE 25984 presented at the SPE/EPA Exploration and Production Environmental Conference, San Antonio, TX, March 7–10, 1993.

Hartmann, S., Ueckert, D. N., and McFarland, M. L., "Evaluation of Leaching and Gypsum for Enhancing Reclamation and Revegetation of Oil Well Reserve Pits in a Semiarid Area," Proceedings of the First International Symposium on Oil and Gas Exploration and Production Waste Management Practices, New Orleans, LA, Sept. 10–13, 1990, pp. 431–441.

Hoffman, J., "A Survey of Regulatory Approaches to Cleanup Standards for Soil Contaminated by Spills of Crude Oil or Salt Water," paper SPE 25985 presented at the SPE/EPA Exploration and Production Environmental Conference, San Antonio, TX, March 7–10, 1993.

Holliday, G. H. and Deuel, L. E., "A Statistical Review of API and EPA Sampling and Analysis of Oil and Gas Field Wastes," paper SPE 20711 presented at the 65th Annual Technical Conference and Exhibition, New Orleans, LA, Sept. 23–25, 1990.

Holliday, G. H. and Deuel, L. E., "Determining Total Petroleum Hydro-carbons in Soil," paper SPE 26394 presented at the 68th Annual Technical Conference and Exhibition, Houston, TX, Oct. 3–6, 1993.

Hunt, J. R., Sitar, N., and Udell, K. S., "Nonaqueous Phase Liquid Transport and Cleanup: 1. Analysis of Mechanisms," *Water Resources Research,* Vol. 24, No. 8, Aug. 1988, pp. 1247–1258.

Johnson, L. D. and James, R. H., "Sampling and Analysis of Hazardous Wastes," in *Standard Handbook of Hazardous Waste Treatment and Disposal.* H. M. Freeman (editor), New York: McGraw-Hill Book Company, 1989.

Johnson, W. B. and Jennings, K. V. B., "Evaluating the Effectiveness of Corrective Actions Involving Groundwater," paper SPE 20062 presented at the 60th California Regional Meeting, Ventura, CA, April 4–6, 1990.

Karisson, H., "Horizontal Systems Technology for Shallow-Site Remedia-tion," *J. Pet. Tech.,* Feb. 1993, pp. 160–165.

Keith, L. H., Crummett, W., Degan, J., Libby, R. A., Taylor, J. K., and Wentler, G., "Principles of Environmental Analysis," *Analytical Chemistry,* Vol. 55, No. 14, Dec. 1983, pp. 2210–2218.

Kiesling, R. W., Alexander, S. K., and Webb, J. W., "Evaluation of Alternative Oil Spill Cleanup Techniques in a *Spartina alterniflora* Salt Marsh," *Environmental Pollution,* Vol. 55, No. 1, 1988, pp. 221–238.

Leggett, S. A., and England, S. L., "Sulphur Block Basepad Reclamation Programs Undertaken at Three Facilities in Central Alberta," Proceedings of the First International Symposium on Oil and Gas Exploration and Production Waste Management Practices, New Orleans, LA, Sept. 10–13, 1990, pp. 945–954.

Maki, A. W., Burns, W. A., and Bence, T. E., "Management of Environmental Impact Studies: A Perspective on the Exxon Valdez Environmental Assess-ment," paper SPE 26677 presented at the 68th Annual Technical Con-ference and Exhibition, Houston, TX, Oct. 3–6, 1993.

National Research Council, *Oil in the Sea: Inputs, Fates, and Effects,* Washington, D.C.: National Academy Press, 1985.

Peters, R. W. and Wentz, C. A., "Remediation of Oil Field Wastes," *Advances in Filtration and Separation Technology, Vol. 3, Pollution Control Tech-nology for Oil and Gas Drilling and Production Operations,* American Filtration Society. Houston: Gulf Publishing Co., 1991, pp. 58–66.

Sims, R. C., "Soil Remediation Techniques at Uncontrolled Hazardous Waste Sites: A Critical Review," Journal of the Air and Waste Management Association Reprint Series: RS-15, 1990.

Testa, S. M., and Winegardner, D. L., *Restoration of Petroleum-Contaminated Aquifers.* Chelsea, Michigan: Lewis Publishers, Inc., 1991.

Udell, K. S. and Stewart, L. D., "Combined Steam Injection and Vacuum Extraction for Aquifer Cleanup," presented at the Conference of the

International Association of Hydrogeologists, Calgary, Alberta, April 18–20, 1990.

Ueckert, D. N., Hartmann, S., and McFarland, M. L., "Evaluation of Containerized Shrub Seedlings for Bioremediation of Oilwell Reserves Pits," Proceedings of the First International Symposium on Oil and Gas Exploration and Production Waste Management Practices, New Orleans, LA, Sept. 10–13, 1990, pp. 403–410.

U. S. Environmental Protection Agency, "International Evaluation of In Situ Biorestoration of Contaminated Soil and Groundwater," EPA 540-90-012, Sept. 1990.

Wojtanowicz, A. K., Field, S. D., Krilov, Z., and Spencer, F. L., "Statistical Assessment and Sampling of Drilling-Fluid Reserve Pits," *SPE Drilling Engineering,* June 1989, pp. 162–170.

Young, G. N., "Guidelines for the Application of Geophysics to Onshore E&P Environmental Studies," paper SPE 23369 presented at the First International Conference on Health, Safety, and Environment, The Hague, Netherlands, Nov. 10–14, 1991.

APPENDIX A

Environmental Regulations

Past environmental practices by segments of the petroleum industry have lead to the loss of public confidence that the industry is able to regulate itself and still protect the environment. Because of this, a large number of environmentally-related laws have been passed, and more are under consideration.

Regulations vary significantly from country to country, state to state, and locality to locality. In most areas, there are multiple, overlapping regulatory agencies that govern various aspects of oil and gas exploration and production. Because these regulations are rapidly changing, any summary of them can be quickly outdated.

Many environmental regulations impose both civil and criminal penalties, with fines and jail terms for violators. Civil penalties can be imposed on both companies and individuals for violations, regardless of intent. Criminal penalties can be imposed on individuals for deliberate violations of the regulations. It is the individual's responsibility to ensure that their actions are in compliance with all existing regulations. The courts in the United States have generally held that supervisors and managers "know" what their employees are doing and thus can be held liable for their employees' actions.

Most environmental laws in the United States are based on the concept of *strict liability*. Strict liability means that neither negligence nor wrongful intent are necessary for liability to be imposed. The company or person that violated the law will be held responsible, no matter what mitigating circumstances may be present, including sabotage or natural disaster.

Good communications between industry, legislators, and regulatory agencies are needed in developing meaningful regulations. Input from industry is important to ensure that new regulations are based on accurate scientific information and that they contribute to real environmental protection without adding a useless burden to industry.

This appendix gives a brief overview of many of the laws and regulations impacting drilling and production activities. More extensive summaries are available (Gilliland, 1993; U.S. Department of Energy, 1991). Regulatory agencies should be contacted prior to initiating any drilling and production activity, however, to ensure that those activities will be conducted in compliance with whatever the current regulations at that time and place may be.

UNITED STATES FEDERAL REGULATIONS

A number of federal environmental regulations affect the upstream petroleum industry. These regulations are complex and require considerable knowledge and effort to ensure compliance. The major regulations are briefly summarized Table A-1 and are discussed below. Additional regulations may also apply that impact drilling and production activities.

Environmental regulations are generally broad and can overlap. In some cases, they can be inconsistent. For example, drilling muds are exempt from the Resource Conservation and Recovery Act (RCRA), Subtitle C, and can be legally disposed of in reserves pits. Reserves pit contents such as drilling muds, however, are not exempt from the Comprehensive Environmental Response, Compensation, and Liability Act (Superfund).

Although the U.S. Environmental Protection Agency is responsible for promulgating these regulations, individual states can be granted primacy if they adopt regulations that are at least as strict as the federal regulations. Most oil and gas producing states have received primacy and these regulations are enforced at the state level.

Resource Conservation and Recovery Act (RCRA)

The Resource Conservation and Recovery Act (RCRA) was initially enacted in 1976 and amended in 1980 to establish a system for managing hazardous solid wastes. This act specifies criteria for determining whether wastes are hazardous or nonhazardous and promulgated requirements on how each are to be managed. Hazardous wastes are

Table A-1
Overview of Federal Environmental Regulations

Resource Conservation and Recovery Act (RCRA)	Regulates management, treatment, and disposal of hazardous wastes.
Safe Drinking Water Act	Regulates injection wells that may contaminate freshwater aquifers.
Clean Water Act	Regulates activities that may pollute surface waters.
Comprehensive Environmental Response, Compensation, and Liability Act (CERCLA)	Regulates cleanup of existing hazardous waste sites.
Superfund Amendments and Reauthorization Act (SARA)	Regulates reporting of storage and use of hazardous chemicals.
Clean Air Act	Regulates activities that emit air pollutants.
Oil Pollution Act	Regulates emergency response plans for oil discharges.
Toxic Substances Control Act	Regulates testing of new chemicals.
Endangered Species Act	Regulates actions that jeopardize endangered or threatened species.
Hazard Communication Standard	Regulates the availability of information on chemical hazards to employees.
National Environmental Policy Act (NEPA)	Regulates actions of federal government that may result in environmental impacts.

regulated under Subtitle C and nonhazardous wastes are regulated under Subtitle D. The regulations for hazardous wastes are considerably more stringent than those of nonhazardous wastes. As discussed below, most but not all wastes generated during drilling and production of oil are exempt from RCRA: Subtitle C.

Under RCRA, a waste is any material that is discarded or is intended to be discarded. It is the intent of future use that determines whether it is considered a waste regulated under RCRA. This act also defines solid wastes as any wastes that are either solid, semisolid, liquid, or gases contained in storage vessels. It further defines a hazardous waste as any solid waste that can cause or significantly contribute to an increase in mortality or in serious irreversible or incapacitating reversible illness, or pose a substantial present or potential hazard to human health or the environment when improperly treated, stored, transported, disposed of, or otherwise managed.

Under RCRA, it is a crime to

1. knowingly cause hazardous materials to be transported to an unpermitted facility or to knowingly transport hazardous materials without a manifest,
2. knowingly treat, store, or dispose of hazardous wastes without a permit or in violation of a permit,
3. knowingly falsify records, labels, manifests, or other documents used for complying with the Act,
4. or knowingly fail to comply with, or interfere with, recordkeeping requirements under the Act.

Violations of RCRA include fines of up to $50,000 per day and two years of imprisonment. If human life is threatened by "knowing endangerment," violations are a crime with fines of up to $1,000,000 and 15 years of imprisonment.

The EPA has established five criteria to determine whether a waste is hazardous or not under this act. There are four generic criteria that are based on the waste properties. These criteria are discussed below. The fifth criterion is for the waste to be listed by name. Listed wastes are those that are known to be hazardous, such as carcinogens and poisons. The designation of whether a material is considered hazardous or not is normally provided on Material Safety Data Sheets.

A waste is considered to be characteristically hazardous if it fits any of the following generic criteria:

- *Ignitability.* A waste is considered ignitable if it presents a fire hazard during routine management. A waste is considered ignitable if it is a liquid and has a flash point less than 140°F; if it is not a liquid and is capable of causing fire through friction, absorption of moisture, or spontaneous chemical changes and, when ignited, burns so vigorously that it creates a hazard; or if it is an ignitable compressed gas or an oxidizer as defined under U.S. Department of Transportation regulations. Examples of ignitable wastes include acetone, isopropanol, hexane, and methanol.

- *Corrosivity.* A waste is considered corrosive if it is able to deteriorate standard containers, damage human tissue, and/or dissolve toxic components of other wastes. An aqueous waste is considered corrosive if it has a pH less than or equal to 2 or greater than or equal to 12.5. A nonaqueous liquid is corrosive if it corrodes SAE 1020 steel at a rate greater than 0.25 inches per year at a temperature of 130°F. Although there is no provision for corrosivity of solids, many states require that a sample be placed in distilled water and the resulting pH be measured. Examples of corrosive wastes include sodium hydroxide, potassium hydroxide, and acids.
- *Reactivity.* A waste is considered reactive if it has a tendency to become chemically unstable under normal management conditions or react violently when exposed to air or mixed with water, or if it can generate toxic gases. Specific regulatory definitions for reactivity have not been developed. Examples of reactive wastes include cyanide or sulfide solutions, water-reactive metals, and picric acid.
- *Toxicity.* A waste is considered toxic if it can leach toxic components in excess of specified regulatory levels upon contact with water. A list of materials and the level above which they would be considered toxic under RCRA is shown in Table A-2. The test procedure to be used, called *toxicity characteristic leaching procedure* (TCLP), is carefully specified under the regulations and is very expensive to conduct. A summary of the TCLP procedure as it applies to the petroleum industry has been prepared by the American Petroleum Institute (American Petroleum Institute, 1990a).

If a waste is considered to be hazardous under RCRA, "cradle-to-grave" management and tracking of the waste is then required, including waste generation, transportation, treatment, storage, and disposal. The generator of the waste can be held liable for the waste, no matter who it has been passed on to or how long ago the waste was disposed.

After an extensive review of wastes generated by the upstream petroleum industry, it was determined that those wastes were not

Table A-2
Regulatory Limits for Toxicity Criterion Under RCRA

Contaminant	Regulatory Level (mg/L)
Arsenic	5.0
Barium	100.0
Benzene	0.5
Cadmium	1.0
Carbon tetrachloride	0.5
Chlordane	0.03
Chlorobenzene	100.0
Chloroform	6.0
Chromium	5.0
o-Cresol	200.0
m-Cresol	200.0
p-Cresol	200.0
Cresol	200.0
2,4-D	10.0
1,4-Dichlorobenzene	7.5
1,2-Dichloroethane	0.5
1,1-Dichloroethylene	0.7
2,4-Dinitrotoluene	0.13
Endrin	0.02
Heptachlor	0.008
Hexachlorobenzene	0.13
Hexachloro-1,3-butadiene	0.5
Hexachloroethane	3.0
Lead	5.0
Lindane	0.4
Mercury	0.2
Methoxychlor	10.0
Methyl ethyl ketone	200.0
Nitrobenzene	2.0
Pentachlorophenol	100.0
Pyridine	5.0
Selenium	1.0
Silver	0.7
Tetrachloroethylene	0.7
Toxaphene	0.5
Trichoroethylene	0.5
2,4,5-Trichlorophenol	400.0
2,4,6-Trichlorophenol	2.0
2,4,5-TP (Silvex)	1.0
Vinyl chloride	0.2

intrinsically hazardous (U.S. Environmental Protection Agency, 1987; American Petroleum Institute, 1983). Because of this, most of these wastes have been exempted from RCRA: Subtitle C. This exemption includes drilling muds, produced water, and other wastes directly associated with drilling and production activities. This exemption gives operators the ability to manage most drilling and production wastes as nonhazardous wastes, although waste management must still be in compliance with the many other existing regulations.

Not all wastes generated during drilling and production are exempt from RCRA. Nonexempt wastes include those that are generated from the maintenance of equipment or that are not unique to exploration and production activities. Furthermore, some exempt wastes can lose their exemption upon custody transfer, e.g., crude oil loses its exemption when it reaches a refinery. Wastes that are sent to certain off-site disposal facilities that are not dedicated to petroleum wastes may also lose their exemption.

A list of RCRA: Subtitle C exempt wastes and a list of RCRA nonexempt wastes are provided in this appendix. A simple rule of thumb can be used to help determine whether a waste is exempt or not. If the waste originated from a well, was introduced into a well, or came into contact with the production stream during removal of produced water or other contaminants from the production stream, the waste is probably exempt.

In addition to the RCRA designation of hazardous wastes, states can also generate their own lists of hazardous and nonhazardous materials. Local regulatory agencies should be consulted for current lists.

Nonexempt wastes are not necessarily hazardous and do not necessarily require management under RCRA: Subtitle C. They are hazardous only if they meet one of the previously mentioned hazardous criteria. If there is reason to believe that a nonexempt waste may exhibit one of the hazardous waste characteristics (toxic, corrosive, ignitable, or reactive), it should be tested to determine whether or not it is hazardous or not.

Mixing of exempt and nonexempt wastes should be avoided, if possible, because the mixture may become nonexempt. The following guidelines can be used to indicate whether or not the mixture would be exempt:

1. Mixing of a *nonhazardous* waste (exempt or nonexempt) with an exempt waste results in a mixture that is also exempt.

RCRA: Subtitle C Exempt Wastes

Produced Water.

Drilling Fluids.

Drill Cuttings.

Rigwash.

Drilling fluids and cuttings from offshore operations disposed of onshore.

Well completion, treatment, and stimulation fluids.

Basic sediment and water and other tank bottoms from storage facilities that hold product and exempt waste.

Accumulated materials like hydrocarbons, solids, sand, and emulsions from production separators, fluid treating vessels, and production impoundments.

Pit sludges and contaminated bottoms from storage or disposal of exempt wastes.

Workover wastes.

Gas plant dehydration wastes, including glycol-based compounds, glycol filters, filter media, backwash, and molecular sieves.

Gas plant sweetening wastes for sulfur removal, including amine, amine filters, amine filter media, backwash, precipitated amine sludge, iron sponge, hydrogen sulfide, scrubber liquids and sludges.

Cooling tower blowdown.

Spent filters, filter media, and backwash (assuming the filter itself is not hazardous and the residue in it is from an exempt waste stream).

Packing fluids.

Produced sand.

Pipe scale, hydrocarbon solids, hydrates, and other deposits removed from piping and equipment prior to transportation. Scale formed in boilers is nonexempt, however.

Hydrocarbon-bearing soil.

Pigging wastes from gathering lines.

(continued on next page)

RCRA: Subtitle C Exempt Wastes
(continued)

Wastes from subsurface gas storage and retrieval, except for the listed nonexempt wastes.

Constituents removed from produced water before it is injected or otherwise disposed of.

Liquid hydrocarbons removed from the production stream but not from oil refining.

Gases removed from the production stream, such as hydrogen sulfide and carbon dioxide, and volatilized hydrocarbons.

Materials ejected from a producing well during the process known as blowdown.

Waste crude from primary field operations and production.

Light organics volatilized from exempt wastes in reserves pits or impoundments or production equipment.

Geothermal production fluids.

Hydrogen sulfide abatement wastes from geothermal energy production.

2. Mixing of a *characteristically hazardous* nonexempt waste with an exempt waste creates a nonexempt hazardous waste if the mixture exhibits the same hazardous characteristic (ignitability, corrosivity, reactivity, or toxicity) as the initial hazardous waste. If the mixture does not exhibit the same hazardous characteristic, the waste is exempt, even if it exhibits a different hazardous characteristic. Testing is required to determine whether the mixture is characteristically hazardous. Mixing of a characteristically hazardous waste with a nonhazardous or exempt waste for the purpose of dilution to make the waste nonhazardous is considered a treatment process and is subject to RCRA: Subtitle C hazardous waste regulations and permitting requirements.
3. Mixing of a *listed hazardous* waste with a nonhazardous exempt waste results in a hazardous nonexempt waste, regardless of the proportions used in the mixture.

RCRA Nonexempt Wastes

Unused fracturing fluids or acids.

Gas plant cooling tower cleaning wastes.

Painting wastes.

Oil and gas service company wastes, such as empty drums, drum rinsate, vacuum truck rinsate, sandblast media, painting wastes, spent solvents, spilled chemicals, and waste acids.

Vacuum truck and drum rinsate from trucks and drums transporting or containing nonexempt waste.

Refinery wastes.

Liquid and solid wastes generated by crude oil and tank bottom reclaimers.

Used equipment lubrication oils.

Waste compressor oil, filters, and blowdown.

Used hydraulic fluids.

Waste solvents.

Waste in transportation pipeline-related pits.

Caustic or acid cleaners.

Boiler cleaning wastes.

Boiler refractory bricks.

Boiler scrubber fluids, sludges, and ash.

Incinerator ash.

Laboratory wastes.

Sanitary wastes.

Pesticide wastes.

Radioactive tracer wastes.

Drums, insulation, and miscellaneous solids.

Even though most drilling and production wastes are exempt, many of the wastes would actually test hazardous by the RCRA criteria (McHugh et al., 1993). The hazardous criteria most commonly responsible for failed tests for ignitability and toxicity, e.g., the benzene concentration of produced water.

Safe Drinking Water Act

The Safe Drinking Water Act was passed in 1974 to protect underground sources of drinking water (USDW) from contamination. USDWs are freshwater aquifers that contain fewer than 10,000 mg/l total dissolved solids or that supply water for human consumption or for any public water system, do not contain minerals or hydrocarbons that are commercially producible, and are situated at a depth or location which makes the recovery of water for drinking economically or technologically practical.

The Safe Drinking Water Act regulates underground injection wells through the Underground Injection Control (UIC) program. This program established five classes of injection wells for different types of wastes:

- *Class I:* Hazardous waste disposal wells and disposal wells for industrial and municipal wastes meeting certain criteria.
- *Class II:* Wells for injecting oilfield fluids, whether for enhanced recovery operations or for disposal and for injecting hydrocarbon liquids into underground storage chambers.
- *Class III:* Wells used for extracting minerals like sulfur, solution mining of minerals, in situ gasification of oil shale and coal, or recovery of geochemical energy to produce electricity.
- *Class IV:* Wells used to dispose of hazardous and radioactive wastes which meet certain criteria.
- *Class V:* Injection wells that do not fall into any of the other four criteria.

Virtually all of the fluids injected into the ground during drilling and production activities use Class II wells. Fluids approved for injection into Class II wells include fluids produced from oil and gas wells, commingled wastewaters from gas plants (if nonhazardous at the time of injection), and fluids injected for enhanced or improved oil recovery operations.

To ensure protection of USDWs, all Class II injection wells require a permit prior to drilling. They must be completed in a zone that is isolated from overlying strata by one or more layers of impermeable zones. The wells must be constructed with quality materials (tubulars and cement) and follow methods to ensure their integrity (the ability to confine fluids to the desired zone). They must be tested every five years for mechanical integrity to verify that they do not provide a flow channel between the injection zone and overlying strata.

Class II wells normally are operated at injection pressures below the fracture pressure of the formation to ensure that vertical fractures are not created that can provide a channel through the impermeable zones to other layers. If it can be shown that any fractures will not extend to USDWs, a permit may be obtained.

A major concern with the operation of Class II injection wells is the presence of nearby wells that may provide a communication path between the (injection) disposal formation and USDWs. When wastewater is injected, the target formation is pressurized. This pressure can drive contaminated water up nearby wells to USDWs. To prevent this from occurring, Class II wells are subject to an "area of review" (AOR) requirement. This requirement states that no wells can exist, within a given area of the Class II well, that are not properly completed or plugged. Normally, this AOR is one quarter of an acre. The AOR requirement may mandate new casing and cement for nearby wells, plugging and abandonment of other wells, and possibly, replugging of previously abandoned wells. If the AOR requirement is not met, a permit for a Class II well may not be granted. If a problem AOR well is located on an adjacent lease, that well must also be fixed before a permit can be obtained.

Clean Water Act

The Federal Water Pollution Control Act Amendments (Clean Water Act) were passed in 1972 to protect surface waters by preventing or minimizing discharges of materials like oil, produced water, or drilling mud. It was amended in 1987 to focus more strongly on toxic discharges and non-point source pollution.

Under this act, the discharge of oil onto surface waters in harmful levels is prohibited without a permit. Surface waters include marine environments, lakes, rivers, ponds, streams, and dry drainage channels that have the potential to flow water. Harmful levels include those that cause a sheen or discoloration on the surface of the water or a sludge or emulsion that can be deposited beneath the water. This act also regulates the discharge of stormwater flowing off a site.

The discharge of pollutants from any point source into surface waters requires National Pollutant Discharge Elimination System (NPDES) or state equivalent permits. For a discharge permit to be obtained for any facility, treatment of the wastes prior to discharge may be required. All discharges of oil into United States surface waters must be reported to the Coast Guard National Response Center in Washington, D.C.

The Clean Water Act is the primary federal regulation governing activities in wetlands (Lesniak, 1994). The act regulates dredging and filling of wetlands, including the construction of access roads and drill pads. Under the current United States policy of "no net loss of wetlands," new wetlands may be required to be created to obtain permits.

The act requires all non-transportation related facilities which have discharged or could reasonably discharge oil into navigable waters to prepare and implement a spill prevention control and countermeasure (SPCC) plan. These plans are required for facilities that have oil storage capacities of more than 660 gallons (16 barrels) in a single tank or 42,000 gallons (1,000 barrels) or more in underground tanks.

SPCC plans are contingency plans for handling potential spills of oil into open waterways. They address drainage around onshore facilities, leak detection and prevention of storage tanks, fluid transport and loading, facility security, pollution prevention systems, and control devices. Each plan must be certified and reviewed every three years by a registered professional engineer. The American Petroleum Institute has prepared a document to assist in preparing SPCC plans (American Petroleum Institute, 1989b).

Under the Clean Water Act, it is a crime to willfully or negligently violate effluent limitations or conditions of a discharge permit. Fines of up to $25,000 per day and one year of imprisonment can be imposed for the first conviction and $50,000 and two years of imprisonment for subsequent convictions. It is also a crime to knowingly violate the requirements of the act or to introduce pollutants or hazardous substances into a public sewer system. Fines of between $5,000 and $50,000 per day and one to three years of imprisonment can be imposed for the first violation and double penalties for subsequent violations. If there is "knowing endangerment" of human health or severe bodily harm, fines of up to $250,000 and 15 years of imprisonment can be imposed on individuals and fines of up to $1,000,000 can be imposed on organizations. Falsifying records or tampering with monitoring devices can result in fines of $10,000 and/or two years of imprisonment for the first conviction and double penalties for subsequent convictions.

Comprehensive Environmental Response, Compensation, and Liability Act (CERCLA)

The Comprehensive Environmental Response, Compensation, and Liability Act (CERCLA, or more commonly known as the Superfund) was passed in 1980 and identifies sites from which releases of hazardous materials might occur or have already occurred. Its purpose is to provide for the cleanup of existing waste sites and to establish a claims procedure for affected parties. Currently, over 700 materials are considered hazardous under CERCLA.

The act identifies *potentially responsible parties* (PRPs) who are associated with each Superfund site. A PRP is anyone that may have contributed wastes to the site, regardless of how much waste was contributed or whether or not the waste was hazardous. A company can also be identified as a PRP if it owned the site at one time, even if it did not dispose of any wastes at the site or if it recently purchased the site and has not conducted any activity on the site.

CERCLA can require any or all PRPs to clean up or pay for the cleanup of the site, without regard to fault. The courts have imposed joint and several liability for cleanup, which can force one PRP to pay for the entire cleanup, even if that PRP contributed only a small amount of nonhazardous wastes to the site. The act also allows for costs of damages to natural resources to be charged to PRPs. Because of the potential for significant future liability under CERCLA, there is a strong economic incentive to properly manage solid wastes, both on-site and off-site.

CERCLA requires most releases of hazardous substances into the environment to be reported unless the release occurs in accordance with a National Pollutant Discharge Elimination System (NPDES) permit granted under the Clean Water Act.

Petroleum products, such as crude oil, crude oil fractions, and some refined products like gasoline, are currently exempt from being considered hazardous wastes under CERCLA. However, other wastes that are exempt from RCRA: Subtitle C may be considered hazardous under CERCLA, including some drilling muds and production chemicals. In fact, several oilfield waste disposal sites that accepted RCRA: Subtitle C exempt wastes have become Superfund (CERCLA) sites because the sites were not managed properly (Fitzpatrick, 1990; Campbell and Akers, 1990).

Under CERCLA, it is a crime to fail to notify the appropriate federal agency of a release of a hazardous substance into the environment and to fail to notify the EPA of the existence of an unpermitted hazardous waste disposal site. Fines of up to $10,000 and one year of imprisonment can be imposed per violation. Penalties of $20,000 and one year of imprisonment can be imposed for knowingly destroying or falsifying records required under CERCLA.

Superfund Amendments and Reauthorization Act (SARA)

In 1986, the Superfund Amendments and Reauthorization Act of 1986 (SARA: Title III) was passed, and added an *emergency planning and community right-to-know* provision to CERCLA. SARA requires owners and operators of facilities that store, use, or release hazardous materials in volumes above a specified threshold to report information about those materials to state and local authorities. This information includes a list of all hazardous chemicals, their volumes, and Material Safety Data Sheets (MSDS). The purpose of this information is to assist local authorities in preparing for emergency responses. SARA also requires releases of these chemicals above a certain amount be reported to the appropriate agencies.

SARA was targeted primarily at industrial sites that maintain large quantities of on-site chemicals over long periods of time. At drilling and production facilities, however, many chemicals, such as drilling or workover chemicals, are present for only a few days a year and not present at any other time. Normal operations make it very difficult to identify the times that specific chemicals are present at any given location.

To simplify the reporting requirements under SARA for the upstream petroleum industry, a *generic hazardous chemical category list* and a *generic inventory* of commonly used chemicals was developed (American Petroleum Institute, 1990b). The generic hazardous chemical category list is a list of chemical *categories* that are typically found on-site at various times. The generic inventory indicates the maximum amount of a material that may be on-site at any given time and how it is stored. A generic hazardous chemical category list, with respective hazards, and a generic inventory can be submitted to local authorities instead of continuously updated lists and hazards of specific on-site chemicals. Generic lists assume that chemicals in all categories are on-site for 365 days per year, regardless of when chemicals are actually present. A total of 65 categories are currently in use.

SARA alters the strict liability of CERCLA by allowing new landowners (potential PRPs) to argue that they are not liable for site cleanup costs because they had no knowledge of the contamination at the time they purchased the land. However, a new landowner must prove that he had made "all appropriate inquiry" into previous ownership. The new owner may still be liable for cleanup if the previous owner cannot be found or has gone out of business.

Clean Air Act

Since the Clean Air Act was initially passed in 1955 to regulate air pollutants to protect human health and the environment, it has been amended a number of times. The most significant amendments were made in 1990. Unlike most other environmental legislation, the 1990 amendments of the Clean Air Act do not set safety standards for pollutant levels; instead, the act requires standards to be set on the *maximum available control technology* (MACT). Thus, the allowed emission levels will be linked with improvements in technology, not safety and health.

Three parts of the 1990 amendments will significantly impact drilling and production activities: Titles 1, 3, and 5. Title 1 addresses emissions in nonattainment areas, i.e., areas that do not meet current air quality standards. Title 3 addresses toxic chemicals and the control technology required. Currently, 189 chemicals are regulated under this act. Title 5 addresses how permits will be granted under the act. The impact of the 1990 amendments will not be clear until the late 1990s.

Most states have established attainment standards for the maximum allowable concentrations of air pollutants in outdoor areas. If the ambient pollutant levels exceed those standards, permits for new facilities that emit air pollutants will be very difficult to obtain. Emission offsets from existing facilities will likely be required for a permit to be obtained.

Under the Clean Air Act, it is a crime to knowingly violate any state-implemented pollution control plan, federal new source performance standards, hazardous air pollution standards, or noncompliance orders. Fines of up to $25,000 per day and one year of imprisonment per violation can be imposed for violations. It is also a crime to knowingly make false statements, representations, or reports, or to tamper with required monitoring devices.

Oil Pollution Act

The Oil Pollution Act was passed in 1990 to expand planning and response activities following an accidental discharge of oil. The act requires a *Facility Response Plan* to be prepared for all facilities that could cause "substantial harm." Facility Response Plans under the Oil Pollution Act differ from SPCC plans under the Clean Water Act in that they address responses after a discharge has occurred, while SPCC plans address the prevention of discharges.

Facility response plans must address emergency notification, equipment and personnel available for response following a discharge, evacuation information, identification and evaluation of previous spills and potential spill hazards, identification of small, medium, and worst-case discharge scenarios and response actions, description of discharge detection procedures and equipment, detailed implementation plans for containment and disposal, training procedures, a description of all security precautions, and diagrams of facilities.

A facility normally is considered capable of causing "substantial harm" if it has a total storage capacity greater than 42,000 gallons (1,000 barrels) and transfers oil over water to or from vessels, or the facility has a total storage capacity greater than one million gallons (23,809 barrels) and meets one of the following conditions: does not have an adequate secondary containment for each storage area, is located where a discharge could cause "injury" to an environmentally sensitive area, is located where a discharge could shut down public drinking-water intake, or has had a reported spill greater than 10,000 gallons (238 barrels) in the previous five years.

Although many drilling and production facilities do not meet these criteria for being capable of causing "substantial harm," it is still prudent for a response plan to be developed. If a spill occurs, having a detailed response plan may limit some liability associated with the spill. Not having developed a response plan could be interpreted by the courts as negligence, even if such plans are not required.

Toxic Substances Control Act

In 1976, the Toxic Substances Control Act (TSCA) was enacted to require the testing of chemical substances and mixtures for assessment of risk to human health or the environment before the substances are manufactured and distributed. This act primarily impacts the chemical and refining industries in their development of new products and processes that require new chemicals. This act may apply to service companies developing improved treatment chemicals.

Under the TSCA, it is a crime to knowingly or willfully violate provisions of the act, use substances that were manufactured, processed, or distributed in violation of the act, or refuse entry or inspection by authorized agents after receiving written notification of a violation of the act. Fines of up to $25,000 and one year of imprisonment can be imposed per violation.

Endangered Species Act

The Endangered Species Act of 1973 prohibits actions that jeopardize endangered or threatened species, including the destruction or modification of the critical habitats used by those species. An endangered species is one that is in danger of extinction throughout all or a significant portion of its range. A threatened species is one that is likely to become endangered within the foreseeable future throughout all or a significant portion of its range. This act has been subsequently amended several times.

The Endangered Species Act has significant implications for the petroleum industry (O'Brien, 1991). If a threatened or endangered species is present at a site, three provisions of the act must be addressed.

First is *interagency consultation,* where the federal agencies that grant permits for the oil and gas industry must consult with the United States Fish and Wildlife Service or the National Marine Fisheries

Service to determine whether the proposed action is likely to jeopardize a threatened or endangered species.

Second is the *taking* provision, where any actions that adversely impact a threatened or endangered species is prohibited. A species is considered to be "taken" if it is harmed, harassed, pursued, hunted, wounded, trapped, captured, collected, or any action is undertaken to conduct those activities. The concept of "harm" includes any actions that significantly disrupt essential behavioral patterns.

Third is an *incidental take* permit, where a low level of "incidental taking" is allowed in exchange for the development of a Habitat Conservation Plan. A Habitat Conservation Plan specifies the impact of the allowed level of taking, steps to minimize or mitigate taking impacts, alternatives considered, and other measures that may be required by the permitting agency.

Violations of the Endangered Species Act can result in a $50,000 fine per offense. Willful violations can result in criminal penalties.

Marine Mammal Protection Act

The Marine Mammal Protection Act of 1972, amended in 1988, prohibits the taking and harassing of marine mammals. This act regulates the use of explosives for removing offshore platforms.

Comprehensive Wetlands Conservation and Management Act

The Comprehensive Wetlands Conservation and Management Act of 1991 provides for management and conservation of wetlands. It regulates activities impacting wetlands.

Hazard Communication Standard

The Hazard Communication Standard (under the U.S. Occupational Safety and Health Administration, or OSHA) requires all employers to identify and list chemical hazards at their facilities. The employers are also required to provide health and safety information about those chemicals and to educate all employees through warning labels, Material Safety Data Sheets, and training programs.

National Environmental Policy Act (NEPA)

The National Environmental Policy Act (NEPA) was adopted in 1969 to ensure that the potential environmental impact from any proposed actions of the federal government or of the private sector that receive federal permits have been considered. This act requires detailed environmental reviews for major actions that may affect the quality of the human environment. These reviews may include extensive environmental impact statements. The impact of actions on threatened and endangered species must be included in the environmental reviews.

STATE REGULATIONS

In addition to the federal regulations discussed above, many states have imposed additional regulations on exploration and production activities for the oil and gas industry. These regulations vary considerably from state to state. A more complete discussion of the regulations of individual states is found in the literature (Interstate Oil Compact Commission, 1990; Boyer, 1990; Crist, 1990; Lynn, 1990; Wascom, 1990; Sarathi, 1991; Smith et al., 1993).

LOCAL REGULATIONS

Local agencies—counties, cities, groups of counties—may also regulate petroleum exploration and production activities. Typical local regulations include those involving noise and dust (particulate) levels at a site. However, air and water pollution, including visual and esthetic impacts, can also be regulated in cooperation with state and federal governments.

REGULATIONS IN OTHER COUNTRIES

Most countries regulate oil and gas activities to minimize their environmental impact. These regulations, however, may be different from those in the United States and can vary considerable from country to country. Many of the regulations of other countries have been discussed by a variety of authors, as indicated in Table A-3.

Table A-3
Discussions of Regulations in Other Countries

Country	Authors
Alberta, Canada	Canadian Petroleum Association (1990)
	Degagne and Remmer (1990)
	Mead and Lillo (1991)
Saskatchewan, Canada	Mutch (1990)
India	Kalra (1990)
Madagascar	Ratsimandresy et al. (1991)
Netherlands	Marquenie et al. (1991)
	Meijer and Krijt (1991)
New Zealand	Hughes (1991)
United Nations	Balkau (1990)

COST OF ENVIRONMENTAL COMPLIANCE

Although many of the environmental regulations have increased the protection of the environment, they have also increased the cost of producing oil. The cost of environmental compliance has been reported to be as high as 10% of the annual expenditures of an oil field (Chappelle et al., 1991). These high environmental costs have encouraged the development of new technologies for waste management that can make waste treatment and recycling more cost effective than simple disposal (Donner and Faucher, 1990).

The potential costs of compliance with RCRA, the Safe Water Drinking Act, the Clean Water Act, and the Clean Air Act are considerable. Initial compliance cost estimates ranged from $15 billion to $79 billion, with additional annual costs of $2 to $7 billion, assuming 1985 levels of industry activity (Godec and Biglarbigi, 1991). Prorating these costs over the current United States production rates gives an approximate incremental cost of environmental compliance of a few dollars per barrel. Not all of these environmentally related costs would be incurred, however, because some recovery operations would become uneconomic and would be terminated. Between 3% and 43% of current production would be lost from environmental regulations with an oil price of $20 per barrel. The development of future reserves was estimated to decrease by up to 42% from the cost of environmental compliance.

In a separate study, the annual costs of environmental compliance for both the upstream and downstream petroleum industry were estimated to range between $15 billion and $23 billion (Perkins, 1991). That study also estimates an approximate cost of environmental compliance of a few dollars per barrel.

If the RCRA: Subtitle C exemption for drilling and production wastes were lost, the cost to the United States petroleum industry has been estimated to be an additional $12 billion annually. This would result in significant reduction in exploration and production activities (U.S. Environmental Protection Agency, 1987).

The cost of environmental compliance must also be considered when selling or purchasing oil and gas properties (Russell, 1989; McNeill et al., 1993). Three areas of major concern are groundwater contamination from production and injection wells or pits; the inability to make property improvements because of construction requirements and regulatory constraints; and failure to comply with existing construction or facilities regulations and failure to conduct monitoring and reporting programs. The ability to obtain the necessary permits to conduct the desired production activities must be assured before a property is purchased.

Liability for CERCLA wastes on a property must also be considered before purchasing the property. To minimize such liability, a staged approach should be conducted to evaluate the potential for the property to contain CERCLA wastes and to evaluate the potential for the site to be declared a superfund site (Curtis and Kirchof, 1993). Detailed and expensive sampling should be considered only if there is a significant potential for hazardous wastes to be found on the property.

REFERENCES

American Petroleum Institute, "Summary and Analysis of API Onshore Drilling Mud and Produced Water Environmental Studies," API Bulletin D19, Washington, D.C., Nov. 1983.

American Petroleum Institute, "Suggested Procedure for Development of Spill Prevention and Control and Countermeasure Plans," API Bulletin D16, Washington, D.C., Aug. 1989b.

American Petroleum Institute, "Applying the Revised Toxicity Characteristic to the Petroleum Industry," Washington, D.C., July 1990a.

American Petroleum Institute, "Bulletin on the Generic Hazardous Chemical Category List and Inventory for the Oil and Gas Exploration and Production Industry," API Bulletin E1, Washington, D.C., July 1990b.

Balkau, F., "International Aspects of Waste Management, and the Role of the United Nations Environmental Programme (UNEP)," Proceedings of the U.S. Environmental Protection Agency's First International Symposium on Oil and Gas Exploration and Production Waste Management Practices, New Orleans, LA, Sept. 10–13, 1990, pp. 543–552.

Boyer, D. G., "State Oil and Gas Agency Environmental Regulatory Programs— How Successful Can They Be?," Proceedings of the U.S. Environmental Protection Agency's First International Symposium on Oil and Gas Exploration and Production Waste Management Practices, New Orleans, LA, Sept. 10–13, 1990, pp. 897–910.

Campbell, R. E. and Akers, R. T., "Characterization and Cleanup of Chromium-Contaminated Soil for Compliance with CERCLA at the Naval Petroleum Reserve No. 1 (Elk Hills): A Case Study," paper SPE 20714 presented at the Society of Petroleum Engineers 65th Annual Technical Conference and Exhibition, New Orleans, LA, Sept. 23–26, 1990.

Canadian Petroleum Association, "Production Waste Management Handbook for the Alberta Petroleum Industry," Dec. 1990.

Chappelle, H. H., Donahoe, R. L., Kato, T. T., and Ordway, H. E., "Environmental Protection and Regulatory Compliance at the Elk Hills Field," paper SPE 22816 presented at the Society of Petroleum Engineers 66th Annual Technical Conference and Exhibition, Dallas, TX, Oct. 6–9, 1991.

Crist, D. R., "Brine Management Practices in Ohio," Proceedings of the U.S. Environmental Protection Agency's First International Symposium on Oil and Gas Exploration and Production Waste Management Practices, New Orleans, LA, Sept. 10–13, 1990, pp. 141–146.

Curtis, B. W., II and Kirchof, C. E., Jr., "Purchase/Sale of Property: The Black Hole of Corporate Liability, Ways to Minimize Risk," paper SPE 25957 presented at the Society of Petroleum Engineers/Environmental Protection Agency's Exploration and Production Environmental Conference, San Antonio, TX, March 7–10, 1993.

Degagne, D. and Remmer, W., "A Practical Approach to Enforcement of Heavy Oily Waste Disposal," Proceedings of the U.S. Environmental Protection Agency's First International Symposium on Oil and Gas Exploration and Production Waste Management Practices, New Orleans, LA, Sept. 10–13, 1990, pp. 783–794.

Donner, C. and Faucher, M., "Recycling of Liquid from Discharged Drilling Waste," presented at the U.S. Environmental Protection Agency's First International Symposium on Oil and Gas Exploration and Production Waste Management Practices, New Orleans, LA, Sept. 10–13, 1990.

Fitzpatrick, M., "Common Misconceptions about RCRA Subtitle C Exemption for Wastes from Crude Oil and Natural Gas Exploration, Development, and Production," Proceedings of the U.S. Environmental Protection Agency's First International Symposium on Oil and Gas Exploration and Production Waste Management Practices, New Orleans, LA, Sept. 10–13, 1990, pp. 169–178.

Gilliland, A., *Environmental Reference Manual for the Oil and Gas Exploration and Producing Industry,* Texas Independent Producers and Royalty Owners Association, Austin, TX, 1993.

Godec, M. L. and Biglarbigi, K., "Economic Effects of Environmental Regulations of Finding and Developing Crude Oil in the U.S.," *J. Pet. Tech.,* Jan. 1991, pp. 72–79.

Hughes, H. R., "Environmental Auditing of Government Administration Systems for the Petroleum Industry," paper SPE 23389 presented at the Society of Petroleum Engineers First International Conference on Health, Safety, and Environment, The Hague, Netherlands, Nov. 10–14, 1991.

Interstate Oil Compact Commission, *EPA/IOCC Study of State Regulation of Oil and Gas Exploration and Production Wastes,* Dec. 1990.

Kalra, G. D., "Regulations and Policy Concerning Oil and Gas Waste Management Practices in India," Proceedings of the U.S. Environmental Protection Agency's First International Symposium on Oil and Gas Exploration and Production Waste Management Practices, New Orleans, LA, Sept. 10–13, 1990, pp. 841–852.

Lesniak, K. Z., "Impact of Wetlands Regulations on Oil and Gas Exploration and Production Activities and Petrochemical Facility Development," Proceedings of Petro-Safe '94, Houston, TX, 1994.

Lynn, J. S., "A Review of State Class II Underground Injection Control Programs," Proceedings of the U.S. Environmental Protection Agency's First International Symposium on Oil and Gas Exploration and Production Waste Management Practices, New Orleans, LA, Sept. 10–13, 1990, pp. 853–864.

Marquenie, J. M., Kamminga, G., Koop, H., and Elferink, T. O., "Onshore Water Disposal in the Netherlands: Environmental and Legal Developments," paper SPE 23320 presented at the Society of Petroleum Engineers First International Conference on Health, Safety, and Environment, The Hague, Netherlands, Nov. 10–14, 1991.

McHugh, B. H., Fox, T. C., and Deans, W. S., "Characteristics of Oil and Gas Production Solid Waste in Montana," paper SPE 25928 presented at the Society of Petroleum Engineers/Environmental Protection Agency's Exploration and Production Environmental Conference, San Antonio, TX, March 7–10, 1993.

McNeill, R. O., Reed, T. M., and Hunnicutt, J. C., "The Importance of Environmental Site Assessments of Oil and Gas Properties Prior to Property

Purchase or Sale: Examples of Environmental Hazards," paper SPE 25954 presented at the Society of Petroleum Engineers/Environmental Protection Agency's Exploration and Production Environmental Conference, San Antonio, TX, March 7–10, 1993.

Mead, D. A. and Lillo, H., "The Alberta Drilling Waste Review Committee— A Cooperative Approach to Development of Environmental Regulations," Proceedings of the U.S. Environmental Protection Agency's First International Symposium on Oil and Gas Exploration and Production Waste Management Practices, New Orleans, LA, Sept. 10–13, 1990, pp. 1–6.

Meijer, K. and Krijt, K., "Implications of The Netherlands' Environmental Policy for Offshore Mining," paper SPE 23339 presented at the Society of Petroleum Engineers First International Conference on Health, Safety, and Environment, The Hague, Netherlands, Nov. 10–14, 1991.

Mutch, G. R. P., "Environmental Protection Planning for Produced Brine Disposal in Southwestern Saskatchewan Natural Gas Fields," Proceedings of the U.S. Environmental Protection Agency's First International Symposium on Oil and Gas Exploration and Production Waste Management Practices, New Orleans, LA, Sept. 10–13, 1990, pp. 375–386.

O'Brien, P. Y., "An Endangered Species Program: The Link Between Compliance and Conservation," paper SPE 23346 presented at the Society of Petroleum Engineers First International Conference on Health, Safety, and Environment, The Hague, Netherlands, Nov. 10–14, 1991.

Perkins, J., "Costs to the Petroleum Industry of Major New and Future Federal Government Environmental Requirements," API Discussion Paper #070, Washington, D.C., Oct. 1991.

Ratsimandresy, R. Raveloson, E. A., and Lalaharisaina, J. V., "Environmental and Petroleum Exploration in Madagascar," paper SPE 23344 presented at the Society of Petroleum Engineers First International Conference on Health, Safety, and Environment, The Hague, Netherlands, Nov. 10–14, 1991.

Russell, R. M., "Environmental Liability Considerations in the Valuation and Appraisal of Producing Oil and Gas Properties," *J. Pet. Tech.*, Jan. 1989, pp. 55–58.

Sarathi, P. S., "Environmental Aspects of Heavy Oil Recovery by Thermal EOR Processes," paper SPE 21768 presented at the Society of Petroleum Engineers Western Regional Meeting, Long Beach, CA, March 20–22, 1991.

Smith, G. E., Smith, W. R., Littleton, D. J., and Simmons, J., "Recent Improvements in State Regulatory Programs and Compliance Practices," paper Society of Petroleum Engineers/Environmental Protection Agency's Exploration and Production Environmental Conference, San Antonio, TX, March 7–10, 1993.

U.S. Department of Energy, "Environmental Regulations Handbook for Enhanced Oil Recovery," NIPER-546, Dec. 1991.

U.S. Environmental Protection Agency, "Management of Wastes from the Exploration, Development, and Production of Crude Oil, Natural Gas, and Geothermal Energy—Executive Summaries," report to Congress, Washington, D.C., Dec. 1987, p. 27.

Wascom, C. D., "A Regulatory History of Commercial Oilfield Waste Disposal in the State of Louisiana," Proceedings of the U.S. Environmental Protection Agency's First International Symposium on Oil and Gas Exploration and Production Waste Management Practices, New Orleans, LA, Sept. 10–13, 1990, pp. 821–832.

APPENDIX B

Sensitive Habitats

Some habitats have a unique sensitivity to oil and gas production activities and require that special operating procedures be followed to minimize impact on them. Of particular concern are rain forests and arctic environments.

RAIN FORESTS

Rain forests provide one of the most biologically diverse environments for oil and gas operations. Because of this diversity, operations in rain forests should be conducted with caution. Drilling muds should be landfilled in dry, lined pits. Formation water should be reinjected, if possible. Onsite treatment of water not reinjected should include aeration for oxygenation and cooling, skimming of surface oil, flocculation and settling to remove solids, and dilution before being discharged into adjacent waterways (Ledec, 1990).

Precautions should be taken to prevent oil spills. Proper spacing of valves and shutoff mechanisms can be used to minimize effects of pipeline leaks. Pipelines should be buried to reduce the risk of vehicles damaging pipelines along roads. Buried pipelines also require less clearing of the forest to maintain a right-of-way along the pipe. Oil storage tanks should have permanent, earthen levees of sufficient size to contain all of the fluids.

Road construction methods should be used to minimize damage to surrounding trees. New camps in forested areas should use the minimum amount of land required for buildings, recreational purposes, and safety. All cleared areas should be rehabilitated when the use is over.

Road construction and the subsequent colonization and conversion of the forest by natives to agricultural uses are a major source of deforestation. These activities can be minimized by reducing the length of roads developed. Rivers and lakes can be used to ship goods to reduce the length of roads needed. Guards along roads can also minimize any illegal travel and limit colonization and logging.

Activities of oilfield personnel should be managed. Sufficient food should be supplied at the site so personnel do not need to supplement their diet locally. Fishing and hunting should be prohibited. Firearms should be prohibited unless security needs demand it. Access of workers to indigenous populations should also be restricted.

ARCTIC REGIONS

Because of the harsh climate, the arctic coast tundra and wetlands are very sensitive and are slow to recover from any disturbances. Limited sunlight, extreme cold, nutrient-poor soils, and permafrost result in low rates of plant growth and excessively prolonged periods of recovery. Because of these conditions, drilling and production activities must be conducted in such a way as to minimize any adverse impact.

One difficulty with operations in the Beaufort Sea is that the shallow waters limit access between the shore and the open sea for transportation of supplies. This problem has been overcome by dredging ports and constructing gravel causeways (Robertson, 1991). Because causeways can alter the natural flow of water in the nearshore region, bridges may need to be constructed to allow channels for water flow.

REFERENCES

Ledec, G., "Minimizing Environmental Problems from Petroleum Exploration and Development in Tropical Forest Areas," Proceedings of the U.S. Environmental Protection Agency's First International Symposium on Oil and Gas Exploration and Production Waste Management Practices, Sept. 10–13, 1990, New Orleans, LA, pp. 591–598.

Robertson, S. B. "Environmental and Permitting Considerations for Causeways Along the Beaufort Sea, Alaska," paper SPE 21764 presented at the Society of Petroleum Engineers Western Regional Meeting, Long Beach, CA, March 20–22, 1991.

Major U.S. Chemical Waste Exchanges

Western Waste Exchange
Arizona State University
Center for Environmental
Studies
Krause Hall
Tempe, AZ 85287

California Waste Exchange
Department of Health Services
Toxic Substances Control
Division
714/744 P Street
Sacramento, CA 95814
(916) 324-1818

World Association for Safe
Transfer and Exchange
130 Freight Street
Waterbury, CT 06702
(203) 574-2463

Southern Waste Information
Exchange
P.O. Box 6437
Tallahassee, FL 32313
(904) 644-5516

Zero Waste Systems
2928 Poplar Street
Oakland, CA 94608
(415) 893-8257

Colorado Waste Exchange
Colorado Association of
Commerce and Industry
1390 Logan Street
Denver, CO 80203
(303) 831-7411

ICM Chemical
20 Cordova Street, Suite 3
St. Augustine, FL 32084
(904) 824-7247

Georgia Waste Exchange
Business Council of Georgia
181 Washington St., S.W.
Atlanta, GA 30303
(404) 223-2264

Industrial Material Exchange
Service
IEPA-DLPC-24
2200 Churchill Road
Springfield, IL 62706
(217) 782-0450

Great Lakes Regional Waste
Exchange
Waste Systems Institute of
Michigan, Inc.
470 Market, SW, Suite 100A
Grand Rapids, MI 49505
(616) 363-7367

Montana Industrial Waste Exchange
P.O. Box 1730
Helena, MT 59624
(217) 782-0450

New Jersey State Waste Exchange
New Jersey Chamber of Commerce
5 Commerce Street
Newark, NJ 07102
(201) 623-7070

Industrial Commodities Bulletin
Enkarn Corporation
P.O. Box 590
Albany, NY 12210
(518) 436-9684

Piedmont Waste Exchange
Urban Institute
University of North Carolina
Charlotte, NC 28223
(704) 597-2307

Louisville Area Industrial Waste
Exchange
Louisville Chamber of
Commerce
1 Riverfront Plaza, 4th Floor
Louisville, KY 40202
(502) 566-5000

Midwest Industrial Waste
Exchange
Rapid Commerce and Growth
Association
10 Broadway
St. Louis, MO 63102
(314) 231-5555

New England Materials
Exchange
34 North Main Street
Farmington, NH 03835
(603) 755-4442 or 755-9962

Alkem
25 Glendale Road
Summit, NJ 07901
(201) 277-0060

Northern Industrial Waste
Exchange
90 Presidential Plaza, Suite 122
Syracuse, NY 13202
(315) 422-6572

Ore Corporation
2415 Woodmere Drive
Cleveland, OH 44106
(216) 371-4869

Techrad Industrial Waste
Exchange
4619 North Santa Fe
Oklahoma City, OK 73118
(405) 528-7016

Tennessee Waste Exchange
Tennessee Manufacturers
Association
501 Union Street, Suite 601
Nashville, TN 37219
(615) 256-5141

Chemical Recycle Information
Program
1100 Milam Building, 25th
Floor
Houston, TX 77002
(713) 658-2462 or 658-2459

Inter-Mountain Waste Exchange
W.S. Hatch Company
643 South 800 West
Woods Cross, UT 84087
(801) 295-5511

Source: B. Quan, "Waste Exchanges," in *Standard Handbook of Hazardous Waste Treatment and Disposal*, H. M. Freeman (editor). New York: McGraw-Hill Book Company, 1989. Used by permission.

APPENDIX D

Offshore Releases of Oil

Perhaps the most obvious environmental impact from drilling and producing oil results from offshore releases of oil. Oil slicks can be carried over large distances and affect many miles of sensitive shorelines. Over time, natural processes will disperse and destroy an oil slick, but often not quickly enough to prevent damage to the shoreline. The best response to offshore releases of oil is to minimize the amount of oil that reaches the shoreline. This can be accomplished by mechanically removing the oil from the water by providing a physical barrier between the oil and shoreline and by enhancing the naturally occurring processes that remove and degrade the oil from the water.

NATURAL DISPERSION OF OIL

When oil is spilled on open water, it is dispersed and destroyed by a number of natural processes. These processes include spreading out over the surface of the water, evaporation of volatile components, dispersion of oil droplets into the water column, attachment of droplets to suspended sediments in the water, dissolution of soluble components into the water column, photo-oxidation of hydrocarbons in the presence of sunlight, hydrolysis, and biological degradation (Jordan and Payne, 1980; National Research Council, 1985). A simplified schematic of these processes is shown in Figure D–1.

When oil is spilled on water, it spreads out over the water surface and moves with the wind and water currents. The thickness of an oil slick is typically between 0.09 and 0.2 mm, with an average thickness of about 0.1 mm (American Petroleum Institute, 1986a). Oil slicks are

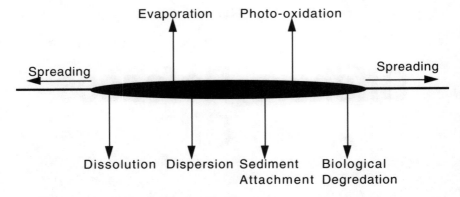

Figure D-1. Dispersion pathways for oil on open water.

not continuous, however; they tend to break up into long patches, with stretches of relatively open water between each patch.

Oil released on open water is transported by local water currents. Because these currents flow parallel to the shoreline, they tend to keep oil slicks away from sensitive shoreline habitats. The motion of oil slicks, however, is also affected by winds, which can blow the slicks to shore. The average speed of a wind-driven oil slick is about 3-4% of the wind speed (National Research Council, 1985).

Following the release of crude oil on open water, evaporation removes between one and two thirds of the oil from the slick during the first few hours (Jordan and Payne, 1980). This evaporation rate, however, depends on the oil composition, temperature, and wind.

Dissolution of hydrocarbon components can also remove some oil from a slick. The solubility of crude oil varies somewhat with composition, but the average solubility is about 30 mg/l (National Research Council, 1985). The most soluble components are the low molecular weight aromatics such as benzene, toluene, and xylene. These compounds, however, are very volatile and are removed primarily by evaporation. Many of the compounds that do dissolve are eventually evaporated back into the air.

Oil slicks can be broken by surface turbulence from wind and wave action into a floating water-in-oil emulsion called chocolate mousse. Mousse, once formed, is long-lasting and very difficult to clean up. The formation of this stable emulsion is more likely for heavy oils at lower temperatures.

Oil that is broken into small droplets can be dispersed into the water column from turbulence as an oil-in-water emulsion. Large droplets will usually float back to the surface and be recombined with the slick. Small droplets, however, can be taken up by marine organisms and incorporated into fecal pellets or can be sorbed onto suspended particles, particularly clays from river runoff. Because the settling rate of suspended particulates can be low, water currents can disperse the sorbed hydrocarbons long distances away from the spill site, keeping their concentration at any particular location relatively low.

Oil that has been either evaporated or dissolved can be decomposed by photo-oxidation when exposed to sunlight. High-energy photons from the sun break the hydrocarbon molecules, which then react with oxygen, destroying the original molecule. The toxicity of partially photo-oxidized hydrocarbons, however, can be higher than that of the original hydrocarbons (National Research Council, 1985). Because the surface-to-volume ratio for an oil slick is low, photo-oxidation does not remove a significant amount of oil from the slick itself.

Some of the dissolved oil compounds can be hydrolyzed. In this process, the normal thermal motion of the molecules in water occasionally breaks a chemical bond on the hydrocarbon. The broken bond then reacts with hydrogen or hydroxyl ions in the water. The reaction can be catalyzed by copper or calcium and can be accelerated if the hydrocarbon is adsorbed onto suspended sediments.

Oil remaining in the marine environment will eventually be removed by biological degradation from bacteria, yeasts, or fungi. The degradation rate, however, depends on the availability of oxygen and nutrients, such as nitrogen and phosphorus. Bacterial degradation is a major mechanism for the eventual removal of hydrocarbons from a marine environment, but is slow compared to other mechanisms.

Degradation rates for oil in the marine environment have been estimated and are summarized in Table D-1 (National Research Council, 1985). Under optimized conditions, degradation can be complete in a few hours to tens of hours. The creation of optimized conditions, however, requires enhancement of virtually all naturally occurring conditions found in nature. Optimized conditions are never found in nature and are virtually impossible to establish outside of the laboratory. If a natural bacterial population has been exposed to hydrocarbons for a prolonged period and has had an opportunity to adjust to their presence (a long incubation period), degradation can be completed in

Table D-1
Biodegradation Rates of Oil in Marine Environment

System	Degradation Rate (g/m³/day)	Degradation Time
Optimized seawater conditions	5–2,500	0.3–144 (hours)
Long incubation period (natural seeps)	0.5–60	0.5–60 (days)
Short incubation period (oil spills)	0.001–0.030	3–82 (years)

Source: National Research Council, 1985.
Copyright © 1985, National Academy of Sciences.
Courtesy of National Academy Press, Washington, D.C.

a few days to tens of days. This condition may be found around some natural seeps. If the hydrocarbons are suddenly added to a bacterial population from an oil spill (a short incubation period), degradation can take years. Because of the very slow degradation rate under oil spill conditions, bacterial degradation is not likely to play a major role in removing oil from slicks.

ENHANCED REMOVAL OF OIL

Because natural removal processes are often too slow to prevent an oil slick from reaching the shoreline, active measures to remove the slick from the water may be required. These processes include mechanically removing the oil from the open water to prevent oil from reaching shorelines and adding materials to the slick to enhance natural removal processes.

Mechanical Methods

Mechanical methods for removing oil from open water normally consist of putting physical barriers between the oil and the shoreline and using skimmers to remove the oil. Physical barriers are normally placed to either concentrate the oil in a small area for easier removal or to keep oil away from very sensitive shoreline habitats.

The most common physical barriers used are floating booms. Booms are vertical sheets that extend above the water level by 4 to 12 inches and below the water level by 12 to 24 inches. Booms come in various sizes for use with different wave heights and wind speeds. For sensitive wetlands with very shallow water, earthen dikes could be constructed as a temporary barrier.

A variety of skimmers are available to mechanically collect oil. Skimmers often use oil-wet sorbent materials like polyurethane or polypropylene to collect the oil. These sorbent materials can absorb many times their weight in oil without collecting much water.

Booms and skimmers are most effective when the waves, wind, and currents are low and when used very soon after the oil has been released. Even under ideal conditions, this equipment is most effective on relatively small spills. In heavy seas or for very large spills, these methods are usually ineffective. Because booms and skimmers are most effective when they are employed very soon after oil has been released, they should be stockpiled near potential release points. A suitable means of rapidly transporting and deploying them is also needed.

Chemical Dispersants

Natural removal processes are accelerated if an oil slick is broken into a large number of smaller droplets. Wind and wave action naturally break up a slick into droplets, but the resulting droplets can easily coalesce back into larger patches of oil. This coalescence can be inhibited by adding chemical dispersants. Most dispersants are surfactants that lower the interfacial tension between the oil and water.

Using dispersants has some important advantages for environmental protection. Oil dispersed into the water column is swept away by the currents and is not easily blown to shore by winds. Dispersants also inhibit the formation of mousse, making the removal of nondispersed oil easier. Dispersants also reduce the tendency of oil to stick to solid surfaces (including suspended particulates, fish eggs, and shoreline rocks), making any subsequent shoreline cleanup easier. Dispersants have also been shown to significantly lower the uptake of oil by suspended sediments (American Petroleum Institute, 1985).

Dispersants, however, do have some disadvantages. They temporarily create a higher concentration of oil in the water column beneath the slick, increasing the impact to biota in the water column. Although some of the older dispersants were toxic, many modern dispersants are less toxic than the oil they disperse. Thus, dispersants increase the short-term impact within the water column, but minimize the long-term impact of oil reaching sensitive shorelines. The short- and long-term environmental impacts of using dispersants must be balanced when considering their use. For spills with little likelihood of reaching sensitive shoreline habitats, the use of dispersants may not be necessary. For spills occurring in deep water that are threatening sensitive shoreline habitats, the use of dispersants may be very beneficial.

A number of field trials of dispersants have been conducted. Dispersants have been found to be effective in accelerating the dissipation of oil slicks and reducing the long-term impact of released oil. The method of application (boat or airplane) and the time the dispersant was applied after the oil release affected the results (American Petroleum Institute, 1986b). For near-shore applications, the use of dispersants was found to lower the uptake of oil by mollusks (American Petroleum Institute, 1986c). In a study on oil released in mangrove, seagrass, and coral reef habitats, dispersed oil was observed to have a lower impact in the intertidal zone than undispersed oil, but it had a higher impact in the subtidal zone (American Petroleum Institute, 1987b).

Dispersants have been applied to several oil slicks, but their results have been inconclusive. Because there was no control during such applications, it has not been possible to determine whether the dispersants actually minimized the environmental impact of the oil.

Dispersants were improperly used on oiled shorelines following the Torrey Canyon tanker accident in 1967. High concentrations of toxic solvent-based cleaners were applied directly to the shoreline to remove the oil. These toxic dispersants severely impacted intertidal organisms and significantly delayed the recovery of the area following the spill. The toxicity of these dispersants, however, resulted more from the aromatic hydrocarbon-based solvents used with the dispersants than from the dispersants themselves.

A number of low-toxicity dispersants have been developed since the Torrey Canyon accident. Bioassays have been conducted on a number of these dispersants and are summarized in Table D-2 (Wells, 1984). By comparing these toxicities with those for various hydrocarbons described in Chapter 3, it can be seen that the toxicity of modern dispersants is considerably lower than that of many hydrocarbons.

To be most effective, dispersants need to be applied within a day or two following the release of oil. However, because of the improper

Table D-2
Toxicity of Dispersants

Species	Dispersant	96-hr LC_{50}*
Invertebrates		
Stony Coral	Shell LTX	162 (1 day)
Ologochaete	Corexit 7664	>1,000
	Finasol OSR-2	>1,000
	Finasol SOR-5	>1,000
Intertidal Limpet	BP1100X	3,700
	BP1100WD	270
Crustaceans		
Amphipods	Various water-based	>10,000
	Various oil-based	200 + 130
Mysids	Various water-based	>4,500
	Various oil-based	150
Brown Shrimp	Various	2,800-10,000 (48 hrs)
Grass Shrimp	Corexit 7664	>10,000 (27°C)
		>100,000 (17°C)
	Atlantic-Pacific	1,000 (27°C)
		1,800 (17°C)
	Gold Crew	150 (27°C)
		380 (17°C)
	Nokomis-3	140 (27°C)
		250 (17°C)
Fish		
Larvae	Corexit 7664	400
Gobies	Shell LT	460
Stickleback	Various water-based	950+ 250
	Various oil-based	10,000
Dace	Various water-based	1,400
Coho Salmon	BP1100X	1,700
Killifish	AP	100 (2 days)

Unless otherwise noted.

Source: after Wells, 1984.

Copyright ASTM. Reprinted with permission.

application of dispersants following the Torrey Canyon accident, getting regulatory approval to use dispersants on oil spills can be difficult to obtain in a timely manner. A detailed contingency plan for the use of dispersants should be developed and submitted to regulatory agencies for review and approval prior to any spill to enhance the likelihood of their being approved after a spill has occurred (American Petroleum Institute, 1987a).

Enhanced Photo-oxidation

Recent studies have shown that photo-oxidation of an oil slick can be significantly enhanced by adding titanium dioxide particles to the slick. Titanium dioxide acts as a catalyst to break the hydrocarbon bonds and accelerate oxidation (Gerischer and Heller, 1991 and 1992).

Bioremediation

Bioremediation has been proposed as a method of accelerating the dispersion of oil slicks on open water. As discussed in Chapter 6, bioremediation of hydrocarbon-contaminated soils can take several months for significant biological degradation of the hydrocarbons to occur, even under optimum conditions. Keeping the optimum combination of bacteria and nutrients in contact with oil on open water for more than a few hours is unlikely. Because of this, bioremediation is not believed to be effective in degrading oil slicks. A test of open-water bioremediation was conducted following the Mega Borg accident (*Oil and Gas Journal*, 1990), but this test was considered inconclusive by most scientists because there was no control.

REFERENCES

American Petroleum Institute, "Surface Chemical Aspects of Oil Spill Sedimentation," API Publication 4380, Washington, D.C., April 1985.

American Petroleum Institute, "The Role of Chemical Dispersants in Oil Spill Control," API Publication 4425, Washington, D.C., Jan. 1986a.

American Petroleum Institute, "The Role of Chemical Dispersants in Oil Spill Control," Washington, D.C., Jan. 1986b.

American Petroleum Institute, "Tidal Area Dispersant Project," API Publication 4440, Washington, D.C., July 1986c.

American Petroleum Institute, "Developing Criteria for Advance Planning for Dispersant Use," API Publication 4450, Washington, D.C., April 1987a.

American Petroleum Institute, "Effects of a Dispersed and Undispersed Crude Oil on Mangroves, Seagrasses, and Corals," API Publication 4460, Washington, D.C., Oct. 1987b.

Gerischer, H. and Heller, A., "The Role of Oxygen in Photooxidation of Organic Molecules on Semiconductor Particles," *J. Phy. Chem.,* Vol. 95, 1991.

Gerischer, H. and Heller, A., "Photocatalytic Oxidation of Organic Molecules at TiO2 Particles by Sunlight in Aerated Water," *J. Electochem. Soc.,* Vol. 139, No. 1, Jan. 1992.

Jordan, R. E. and Payne, J. R., "Fate and Weathering of Petroleum Spilled in the Marine Environment: A Literature Review and Synopsis," Ann Arbor Science Publishers, Ann Arbor, MI, 1980.

Oil and Gas Journal, Aug. 6, 1990.

National Research Council, *Oil in the Sea: Inputs, Fates, and Effects,* Washington, D.C.: National Academy Press, 1985.

Wells, P. G., "The Toxicity of Oil Spill Dispersants to Marine Organisms: A Current Perspective," in *Oil Spill Chemical Dispersants: Research, Experience, and Recommendations,* T. E. Allen (editor), STP 840, American Society for Testing and Materials, Philadelphia, PA, 1984.

Index

N

National Pollutant Discharge
Elimination System, 115,
242, 243
Natural gas, 51–52, 57
Neutralization, 185
Nitrogen dioxide, 57, 195
NORM, 6, 56, 126, 146, 211
Nuclear radiation, 54–57, 121–126

O

Offshore platforms, 128, 211
Oil slicks, 261
Oxidation, 180, 195
Oxygen depletion, 42

P

Paraffin inhibitors, 106
Particulates, 196
Percolation, 8, 186, 204
pH, 25, 49, 140, 185, 234
Photo-oxidation, 263, 268
Pipe dope, 30
Plate separators, 174
Precipitation, 180, 183
Produced water, 152
  hydrocarbons, 41
  metals, 41
  process, 39
Production chemicals, 43
  toxicity, 105–106
Profile modification, 48
Pump and treat, 222
Pyrolysis, 189

R

Radioactive decay, 121

Radioactive tracers, 55
Rain forests, 256–257
Recycling, 161–162
Regulations, 10, 230, 249
  Clean Air Act, 245–246
  Clean Water Act, 241–243
  Comprehensive Environmental
    Response, Compensation,
    and Liability Act, 243–244
  Comprehensive Wetlands
    Conservation and
    Management Act, 248
  Endangered Species Act, 247–248
  Hazard Communication
    Standard, 248
  Marine Mammal Protection
    Act, 248
  National Environmental Policy
    Act, 249
  Oil Pollution Act, 246–247
  reserves pits, 38
  Resource Conservation and
    Recovery Act, 149, 231–240
  Safe Drinking Water Act, 240–241
  Superfund Ammendments and
    Reauthorization Act, 244–245
  Toxic Substances Control Act,
    247
Reinjection, 52, 156, 207
Remediation, 9, 64, 216, 220
Reserves pits, 35, 119, 154,
  186–187, 210, 225
Reverse osmosis, 184
Risk assessment, 8, 128–131, 217
Road spreading, 207

S

Salt, 32, 119, 182, 204, 205,
  207, 210, 225

toxicity, 5, 96–100
Sand, 51, 53
Scale, 44, 56
Scrubbers, 52–53, 195, 196
Segregation, 8, 153
Separations, 8, 33, 39, 51, 53,
    160, 172–173, 181
Site assessment, 216
Site preparation, 38, 153, 205,
    256
Solidification, 193
Solvents, 48, 190, 192, 266
Spill prevention control and
    countermeasure plans, 242
Steam injection, 53, 58, 194, 223
Substitution, 29, 156–159
Sulfur, 225
Sulfur dioxide, 44, 53, 57, 127,
    195
Supercritical fluids, 191
Surfactants, 26, 29, 40, 43–49,
    106, 157, 188, 192, 222, 265

T

Toxicity, 4, 5, 71, 234, 263, 266
    air pollution, 126–127
    drilling fluids, 6, 106–120
    heavy metals, 6, 100–105

hydrocarbons, 83–96
nuclear radiation, 121,
    123–126
produced water, 120–121
production chemicals, 105–106
salt, 5, 96–100
Training, 165
Treatment, 8, 162, 172

U

Ultraviolet irradiation, 180

V

Viscosity, 21, 28, 49, 50
Vitrification, 194
Volatile organic carbon (VOC),
    57, 65, 161, 179, 194, 206,
    216, 223
Volatilization, 179, 223

W

Washing, 188, 222
Waste management plans, 7, 144,
    149
Waste migration, 139, 221
Waste minimization, 150–161
Water vapor, 51
Wettability, 48